Makers of Western Science

Makers of Western Science

The Works and Words of 24 Visionaries from Copernicus to Watson and Crick

TODD TIMMONS

McFarland & Company, Inc., Publishers
Jefferson, North Carolina, and London

ISBN 978-0-7864-6061-8
softcover : acid free paper ∞

LIBRARY OF CONGRESS CATALOGUING-IN-PUBLICATION DATA

British Library cataloguing data are available

Front cover photographs: *from top to bottom right* painting of Galileo Galilei
by Justus Susterman, 1636; Albert Einstein, 1921; Charles Darwin, 1854; Niels Bohr,
1885; J. Robert Oppenheimer, 1904; Francis Galton, 1822 (all Wikipedia);
cover design by David K. Landis (Shake It Loose Graphics)

Manufactured in the United States of America

*McFarland & Company, Inc., Publishers
Box 611, Jefferson, North Carolina 28640
www.mcfarlandpub.com*

Table of Contents

Acknowledgments

Many thanks to all the people who helped me with the process of researching and writing this book: Wilma Cunningham, Martha Coleman, and all the fine library staff at the University of Arkansas–Fort Smith Boreham Library; Bruce Bradley, Cindy Rogers and the staff at the Linda Hall Library; Dr. Jim Christiansen and Dr. Amy Skypala who provided input on two particularly difficult areas of the book; and especially my wife, Becky, who served as sounding board, editor, and technical advisor through the entire process.

Preface

The perception that most non-scientists have of science is that it is dry and boring and pursued by dry and boring people. This is an unfortunate stereotype. Throughout history, science has been practiced by passionate, interesting, and even controversial personalities whose biographies shed light on human society and culture. From Galileo's difficulties with the Inquisition, to the quirkiness of Newton, to the iconic figure that was Einstein, science has been the product of fascinating people seeking to explain the world around them.

Science is, after all, about the way we understand our world. It is about theories and grand ideas; it is about instruments and laboratories; it is about connections and themes; but mostly science is about people. Science is one of the intellectual pursuits that make humans unique, and the people who made modern science form the basis of some of the most fascinating tales in human history. The purpose of this book is to chronicle the history of science through the words of the scientists themselves. Along the way, their words also illuminate the human side of science. Each chapter of this book is centered on a particular scientist or group of scientists: part biography, part history, and all about the human endeavor we call science.

We as humans are interested in the lives of other humans. We are interested in the politician as well as the politics; in the poet as well as the poetry; even in the athlete as well as the athletic contest. Events are important, but human interest is pivotal. Natural selection is interesting — and understanding it is important for an educated person — but following Darwin's decades-long struggle to come to terms with mutability of the species is fascinating at a more human level. Understanding the importance of how the moons of Jupiter supported the heliocentric model is valuable, but following Galileo's nightly diary as he realizes that they are not stars, but moons, is captivating. Knowing Kepler's first law of planetary motion is important, but understanding his "Eureka" moment of understanding is more revealing. And Oppenheimer's reaction upon witnessing the explosion of the first atomic bomb says more about the true meaning of this terrible new weapon than all the technical understanding of nuclear fission possibly could.

There are many fine books that trace the history of science through the ages; there are also anthologies of science writing with extended passages from some of the great works in the history of science. This book attempts to combine the two by relating the history of modern science to the reader using extensive passages from the works of the scientists themselves. Who better to appeal to our common sense concerning the truth of a sun-centered universe than Copernicus himself? We let Kepler express in his own words the way in which he awoke to the revelation of elliptical orbits. And Darwin, traveling and observing spectacular sights during his five-year voyage on the *Beagle*, shares with us his slowly evolving ideas leading to the theory of natural selection.

Another advantage to reading the original words of these scientists is that a little of their personality often comes through. Kepler, for instance, is often verbose and tedious: yet, the reader cannot help but sense his excitement when he relates the instant that his revolutionary idea about the orbits of the planets finally comes to him. Newton, usually straightforward and to the point, reveals a not so dispassionate side to his personality when he deals with other scientists he feels have wronged him. And Darwin — whose Victorian prose makes his scientific writing seem like poetry — allows us a peek into his personal life as he weighs the pros and cons of marriage.

Although each chapter focuses on the work of a particular scientist (or group of scientists), broader themes in the history of science are also explored. For instance, Chapter 4 focuses on the contributions to the emergence of the experimental philosophy by William Harvey and William Gilbert, but it also addresses the role of other contributors to this important movement, along with the rise of the scientific societies, and the influence of experimental philosophy on scientific thought and scientific practice. Similarly, Chapter 8 is built around the "father of modern chemistry," Antoine Lavoisier, but the account of the beginning of modern chemistry would not be complete without the contributions of many other chemists (and alchemists and iatrochemists and other proto-chemists) who contributed to the new science of chemistry. And, no coverage of Lavoisier would be complete without mention of the sudden and unfortunate ending to his illustrious life.

Although this book does not attempt to relate the entire history of modern science (a next to impossible undertaking given the sheer amount of history such a work would encompass), it does attempt to convey to the reader some of the most important episodes, events, and especially personalities, that guided — and sometimes misguided — the development of science.

Science in the Ancient and Medieval World

Why begin a book on the history of modern science with an overview of the history of ancient science? For the same reasons one does not commence a study of World War II with Germany's invasion of Poland. To begin in the year 1939 would not do justice to the important events in the years preceding the actual outbreak of hostilities. What were the political, social, and economic reasons for the war? How can one possibly understand the causes of World War II without at least a cursory understanding of the rise to power by the Nazis? And the rise of the Nazis must be understood in the context of the conditions in Germany after World War I. Similarly, the debate over America's entrance into the war makes sense only in the framework of an understanding of isolationist feelings in the United States and policies of previous administrations. To use another analogy, how could one really understand the American Revolution if the starting point is taken as July 4, 1776? Without the context of the Boston Massacre, the Boston Tea Party, the Stamp Act, and so on, what happens in the years to come makes little sense.

For the same reasons, it is impossible to understand modern science without at least a basic knowledge of science before the modern era. Even the very term modern science is problematic — modern may be defined in so many ways. For this book, modern science begins with the work of Copernicus in the sixteenth century, an episode that initiates what is often called the scientific revolution. But what was science revolting against? This brief introduction will attempt to set the stage for modern science with a very concise overview of science before the sixteenth century. For a deeper look into ancient and medieval science, the reader is referred to the list of sources at the end of this introduction.

When did science begin? The question itself is fraught with many difficulties, not the least of which is the definition of science. What passed for science in the ancient world, or in classical Greece, or in medieval Islamic cultures, looks very different from most modern conceptions of science. Therefore, we must meet ancient science on its own terms, withholding judgment based on modern standards. Who were the most influential scientists of the pre-modern world? Again, problems abound in answering such a question. The word scientist was not even coined until the nineteenth century. So even when we use the word scientist to describe someone like Isaac Newton (much less Aristotle) we are using the word anachronistically. Traditionally, those people who searched for truths in the natural world — what we might call a scientist today — were referred to as natural philosophers. Therefore, when we search for the origins of science and for the first scientists we must be careful to understand these terms either did not exist or may have had different meanings hundreds, or even thousands, of years ago.

Perhaps the earliest cultures that pursued disciplines we might call science today were the Babylonians and the Egyptians. At various times in the first and second millennia B.C.E., learned individuals in both cultures developed ideas in basic mathematics like geometry and arithmetic; observational and mathematical astronomy; and technologies like metallurgy, agriculture, spinning, and weaving. However, like nearly all ancient cultures, the Babylonians and the Egyptians traditionally reverted to an explanation of the natural world steeped in magic, mysticism, and the occult. Their world was controlled by the gods, not by natural law.

It was left to the ancient Greeks, working over an extended time period (roughly the sixth century B.C.E. to the second century C.E.) to make the first organized attempts at a systematic explanation of natural phenomena (mostly) stripped of mystical and magical underpinnings. The first rational philosophers in Greece speculated on such things as the shape of the earth, the composition of matter, and the nature of the universe. The Pythagoreans, a cult mixing rational thought with number mysticism, emphasized the mathematical description of phenomena. The famous Greek philosopher Plato argued for the role of the senses and the mind in science. It was Plato who popularized the notion that the world was composed of four fundamental elements: fire, air, water, and earth.

Another Greek philosopher and mathematician, Eudoxus, modeled planetary motion with a series of nested spheres with the earth at the center. Eudoxus' system required 25 connecting spheres to model the movement of the planets, with an additional sphere representing the fixed stars on the outer edge of the universe. Many other Greek thinkers, from Archimedes to Zeno, made significant contributions to natural philosophy; however, the greatest of the ancient Greek philosophers, Aristotle, stands out not only for his impressive body of work but also for the influence that work wielded over western civilization for thousands of years.

Aristotle (384–322 B.C.E.), student of Plato, tutor to Alexander the Great, and founder of perhaps the most famous school of the ancient world (the Lyceum), was perhaps the most influential natural philosopher (scientist) in history. Although it is uncertain just how many books he wrote (probably in the hundreds), the ones that do survive were used as the blueprint for scientific thought throughout the Western world for 2000 years. He wrote on the natural sciences, the physical sciences, astronomy, physiology, and much, much more.

There are several reasons Aristotle's scientific output dominated thought for millennia. One was the adoption of the Aristotelian corpus by medieval Islamic scientists who translated, commented on, corrected, and added to Aristotle's many works. The second reason for the remarkable longevity of Aristotelian science was the support of the Catholic

Church. Aristotle's cosmology placed the earth at the center of the universe; moreover, Aristotle's physics was predicated upon this assumption. Motion was due to objects seeking their natural place in the world, and since Earth was the center of the universe heavy objects — those composed primarily of the element earth — sought their natural place at the center. These views meshed well with the church's teachings that humans were the center of God's creation. Under church guidance, Aristotelianism became orthodox and integrated into the accepted world view.

The basic tenets of Aristotle's cosmology were incorporated by other Greek natural philosophers into mathematical models used to predict the movement of the planets. This mathematical astronomy culminated with the work of Claudius Ptolemy. Ptolemy (85–165) was a Greek mathematician, astronomer, and geographer who thrived at Alexandria, Egypt, the center of learning for the Greek empire. The Library at Alexandria housed the greatest collection of manuscripts in the ancient world. The Library also functioned as a sort of early university, with gardens, museums, lecture halls, and meeting rooms where scholars congregated and students flocked from all over the known world.

Although Ptolemy wrote major works on mathematics, geography, and optics, his most important contribution to ancient science was his astronomical treatise known today as the *Almagest*. Originally entitled *The Mathematical Compilation*, Ptolemy's masterpiece was later referred to as *The Greatest Compilation*. Translated into Arabic and its title shortened, the treatise became simply known as the *Almagest* (the *Greatest*). This work became the final word in astronomy for fifteen centuries.

The *Almagest* presented a geocentric world — the earth at the center of the cosmos with the moon, the sun, the planets, and the stars revolving on spheres around the center. However, to make the observational data fit the mathematical theory, Ptolemy's universe had complications. There were epicycles, or circles upon circles, to explain the positions of the planets. The epicycles helped to model the apparent change in speed of the planets as seen from the earth. Epicycles also explained the phenomena of retrograde motion — when a planet appeared to slow down and reverse directions in its path across the sky. Ptolemy also employed an eccentric — a shift of the center of the universe so that the planets and stars revolved around a point slightly away from the actual earth. Even these two concessions to reality did not fully solve the problem of making the model fit the observations. So Ptolemy added a third geometric construction to his model, the equant. The equant was a point not at the center of the universe or at the place occupied by the earth, but rather at another point near the center from which the apparent motion of the planets was uniform. All in all Ptolemy's system was a marvelously accurate and mathematically sophisticated — if rather messy — model of how the universe operated. Ptolemy was one of the last of the great philosophers to thrive before the fall of the Roman Empire.

The period from the fifth to the fifteenth century is called by several different names: the Middle Ages, the Medieval Period, or the Dark Ages. The Dark Ages is an unfortunate moniker. While Europe struggled through this long period of intellectual stagnation, other parts of the world — particularly the Islamic controlled areas of the Middle East, Northern Africa, and parts of Southern Europe — flourished. Islamic scientists translated many of the writings of the Greeks, and in numerous cases preserved the only copies whose existence survived to modern times. More than simply acting as repositories of

Greek works, Islamic intellectuals added to the store of knowledge and in many cases integrated Greek learning with that of other cultures to make important advances in science, mathematics, and other disciplines. For instance, the ninth-century Persian scholar, Abu Ja'far Muhammad ibn Musa Al-Khwarizmi, combined elements of Babylonian, Indian, and Greek mathematics in his work *The Condensed Book on the Calculation of al-Jabr and al-Muqubala*— considered the first book of algebra (the word *al-Jabr* in the title gives us the term algebra).

The works of the Greek masters, some in the original Greek but most in their Arabic translations, began to trickle into Europe during the twelfth century. The trickle soon became a flood and the revival of Greek philosophy helped to trigger the Renaissance. European universities, first established in Bologna, Salerno, Oxford, Cambridge, and Paris in the eleventh and twelfth centuries, began assimilating these manuscripts, especially those of Aristotle, into their curriculum. Translated into Latin, Aristotle's works dominated European universities for several centuries. Gutenberg's introduction of the printing press in the mid–fifteenth century meant that the Greek corpus, along with a flood of new tomes by emerging European intellectuals, could become accessible to more of the reading public. Italy was the center of the Renaissance, and it was during these exciting times that a young Polish man named Nicolaus Copernicus traveled with his brother to the University of Bologna to further their theological studies.

SOURCES

Bertman, Stephen. *The Genesis of Science: The Story of Greek Imagination.* Amherst, NY: Prometheus Books, 2010.

Clagett, Marshall. *Greek Science in Antiquity.* New York: Abelard-Schuman, 1955.

Lindberg, David. *The Beginnings of Western Science: The European Scientific Tradition in Philosophical, Religious, and Institutional Context, 600 B.C. to A.D. 1450.* Chicago: University of Chicago Press, 1992.

_____. *Science in the Middle Ages.* Chicago: University of Chicago Press, 1978.

Lloyd, G. E. R. *Magic, Reason, and Experience: Studies in the Origin and Development of Greek Science.* Cambridge and New York: Cambridge University Press, 1979.

Rihll, T. E. *Greek Science.* Oxford and New York: Oxford University Press, 1999.

Saliba, George. *Islamic Science and the Making of the European Renaissance.* Cambridge: MIT Press, 2007.

CHAPTER 1

The Copernican Revolution

Therefore, having obtained the opportunity from these sources, I too began to consider the mobility of the earth. And even though the idea seemed absurd, nevertheless I knew that others before me had been granted the freedom to imagine any circles whatever for the purpose of explaining the heavenly phenomena. Hence I thought that I too would be readily permitted to ascertain whether explanations sounder than those of my predecessors could be found for the revolution of the celestial spheres on the assumption of some motion of the earth.—Nicolaus Copernicus, *On the Revolutions of the Heavenly Spheres* (1543)

Before the sixteenth century, the consensus view was that the earth was the center of the cosmos. In fact, very few would even consider that an alternative existed. Everyday experience and common sense, as the argument went, tell us that the earth is at the center of everything. Aristotle, Eudoxus, Ptolemy, and countless other Greek philosophers tell us the same thing. The philosophers in the great modern universities of Europe, the church, the church fathers — even the Bible — were unified in maintaining the same geocentric belief. No sane, reasonable person, the argument concluded, could hold any doctrine except the earth is the center of the cosmos.

Such was the ingrained mindset of sixteenth-century Europe — the mindset that a Polish church official by the name of Nicolaus Copernicus sought to change and in the process began a revolution that would permeate almost every aspect of human thought.

Early Life of Copernicus

Nicolaus Copernicus (the Latinized form of his name, taken when he became a university student; originally he was known as Mikolaj Kopernik, Niclas Kopernik, or Nicolaus Koppernigk) was born in Torun, Poland, on February 19, 1473 — almost twenty years before Columbus embarked on his famous voyage. His father, a copper merchant, migrated from Krakow to Torun where he became a community leader and married the daughter of a well-to-do church official. Nicolaus was the youngest of four children, and received a comfortable — if not privileged — upbringing. His father died when Nicolaus was but ten years old, so his uncle, Lucas Watzenrode, took charge of Nicolaus and his siblings. Uncle Lucas, who was a canon in the cathedral at Frauenburg (and later bishop of Ermland), saw to it that Nicolaus and his brother received an excellent education. This education included further elementary training in his hometown of Torun followed by a more intense humanistic education at a cathedral school. The Copernicus brothers matriculated to the University of Krakow, in the then Polish capital, where it was assumed they would train for a career in service to the church.

The curriculum at Krakow was standard for the times. Copernicus studied Latin classics, mathematics — especially as applied to astronomy and perspective — and the philosophy of Aristotle and his commentators. The astronomy taught at Krakow was of course Ptolemaic astronomy. The emphasis was on practical applications of mathematical astronomy: calendar making for accurate calculations of holy days and observances; terrestrial surveying and navigation skills; and the practical aspects of astrology, such as casting horoscopes. Young Copernicus developed an interest in mathematical astronomy while at Krakow, as evidenced by some of the books he purchased. These include important tables used for astronomical calculations as well as a copy of Euclid's classic work in geometry, *The Elements*.

Copernicus left Krakow without obtaining a degree, something not uncommon for the time. After a brief stay back in Torun, Uncle Lucas sent Copernicus and his brother to study canon law at the University of Bologna. While in Italy, a very important event guaranteed Copernicus a secure future in which he could pursue his studies without excessive financial worries. Once again Uncle Lucas — by this time bishop of Ermland — stepped in, securing for both of his nephews the positions of canon of Frauenburg. These positions included generous benefits without great demands; in fact, both Copernicus brothers continued as canons (with salary) in absentia as they studied at Bologna. Without such support, Copernicus would have found it very difficult throughout his life to maintain the program of study he sought. Patronage, whether from wealthy individuals or from a powerful organization such as the Catholic Church, was extremely important in a time when science was not a profession from which a talented young man could earn a living.

Two very fortuitous circumstances surrounding Copernicus' stay in Bologna conspired to shape the budding astronomer's future. The first was his association with Domenico Maria de Novara, an eminent astronomer who may have planted the seeds of doubt concerning the Ptolemaic system into his students' minds. It was also at Bologna that Copernicus came into direct contact with the humanist movement sweeping through Europe. Humanists emphasized the study of ancient works, especially those Greek works making their way back into Europe after centuries of neglect. Copernicus' mastery of Greek proved to be important in his study of astronomy, especially Ptolemy.

While in Bologna, Copernicus began making astronomical observations, continuing this practice during a yearlong stay in Rome where he also gave lectures in astronomy. Copernicus left Bologna in 1501, once again without a degree. For the next two years, Copernicus studied medicine at Padua, and then went to Ferrara where he finally was awarded a doctorate in canon law. After these extensive travels through the center of the

Italian Renaissance, Copernicus returned to his native land where he spent the rest of his life.

Upon returning to his homeland, Copernicus took up residence in Heilsburg, where he attended to his sick Uncle Lucas and also served as his uncle's secretary and assistant. It also appears that during this time (1506–1512) Copernicus did not actively pursue astronomical studies, spending his time instead assisting his uncle, occasionally practicing medicine, and immersing himself in the affairs of the church. With his uncle's death in 1512, Copernicus resumed his position as canon at Frauenburg and continued his duties for the church. Throughout the rest of his life, Copernicus would juggle these official duties with his interest in astronomy, the result being only one publication in the field in which he would later become known around the world.

Copernicus' Astronomy

Luckily for Copernicus — and for posterity — his position allowed him the resources — if not always the time he desired — to pursue his studies in astronomy. The castle towers at Frauenburg became Copernicus' private observatory. By 1514 he had advanced in his observations and calculations to the point that he circulated a tract known as the *Commentariolus* (*Little Commentary*). In this handwritten manuscript, distributed to astronomers throughout Europe, Copernicus laid out seven underlying assumptions:

1. There is no one center of all the celestial circles or spheres.
2. The center of the earth is not the center of the universe, but only of gravity and of the lunar sphere.
3. All the spheres revolve about the sun as their mid-point, and therefore the sun is the center of the universe.
4. The ratio of the earth's distance from the sun to the height of the firmament is so much smaller than the ratio of the earth's radius to its distance from the sun that the distance from the earth to the sun is imperceptible in comparison with the height of the firmament.
5. Whatever motion appears in the firmament arises not from any motion of the firmament, but from the earth's motion. The earth together with its circumjacent elements performs a complete rotation on its fixed poles in a daily motion, while the firmament and highest heaven abide unchanged.
6. What appear to us as motions of the sun arise not from its motion but from the motion of the earth and our sphere, with which we revolve about the sun like any other planet. The earth has, then, more than one motion.
7. The apparent retrograde and direct motion of the planets arises not from their motion but from the earth's. The motion of the earth alone, therefore, suffices to explain so many apparent inequalities in the heavens.

These seven assumptions form the foundation of the Copernican universe. Note the revolutionary implications, yet stark simplicity, of each statement.

"There is no one center of all the celestial circles or spheres." For millennia, the accepted model of the cosmos had every heavenly body revolving around the earth. The Copernican system did more than simply move the sun to the center of the universe. By having the planets (including the earth) revolve around the sun, while the moon revolved around the earth, Copernicus changed completely the concept of center. There was, he claimed, more than one center of revolution.

"The center of the earth is not the center of the universe, but only of gravity and of the lunar sphere." Gravity here does not have the same definition as in modern physics. It simply refers to the property of heaviness that explains why things tend towards the earth.

"All the spheres revolve about the sun as their mid-point, and therefore the sun is the center of the universe." Copernicus made a simple and innocuous swap: the sun for the earth.

"The ratio of the earth's distance from the sun to the height of the firmament is so much smaller than the ratio of the earth's radius to its distance from the sun that the distance from the earth to the sun is imperceptible in comparison with the height of the firmament." Copernicus realized that one of the implications of his model was that the universe must be *much* larger than previously imagined. The height of the firmament (distance to the fixed stars) was such that it dwarfed the planetary distances.

"Whatever motion appears in the firmament arises not from any motion of the firmament, but from the earth's motion. The earth together with its circumjacent elements performs a complete rotation on its fixed poles in a daily motion, while the firmament and highest heaven abide unchanged." The obvious explanation of the appearance of movement of the stars through the night sky was that the stars revolved around the earth. Copernicus offered another explanation. This apparent motion was actually caused by the earth spinning on its axis.

"What appear to us as motions of the sun arise not from its motion but from the motion of the earth and our sphere, with which we revolve about the sun like any other planet." Furthermore, continued Copernicus, the apparent movement of the sun on the ecliptic is not caused by the sun's movement at all; rather, this motion can be explained by the earth's yearly revolution around the sun. Notice also, Copernicus made the inflammatory statement that the earth is *"like any other planet."*

"The apparent retrograde and direct motion of the planets arises not from their motion but from the earth's. The motion of the earth alone, therefore, suffices to explain so many apparent inequalities in the heavens." Copernicus found beauty in the fact that his system simply and elegantly explained certain inequalities in the movements of the heavenly bodies — in particular retrograde motions. He needed no ad hoc explanations like epicycles to explain these motions (although in *On the Revolutions* we do find that Copernicus requires epicycles to help explain other anomalies in the planetary motions).

Copernicus presented no mathematical proof of his ideas, stating that he reserved such proofs for a later, more exhaustive work. And it is not clear how many read the *Commentariolus*, or exactly how influential it was. However, it is known that an Austrian scholar, Albert Widmanstadt, lectured on the Copernican system in 1533 before a group of church officials which included Pope Clement VII. Then in 1536 Cardinal Schönberg wrote a letter to Copernicus urging him "to communicate this discovery of yours to scholars, and at the earliest possible moment to send me your writings on the sphere of the universe." Although these events seem encouraging, Copernicus delayed publication of his completed work for another decade.

Why did Copernicus wait to publish the full mathematical treatment of his theory? Fear of criticism and rejection certainly played a part. However, duties to church and country also kept Copernicus busy with other tasks. In addition to the day-to-day duties of his office — record keeping, meetings, and other responsibilities of a canon — Copernicus at times held other administrative positions in the church that required his attention, sometimes for years at a stretch. At various times diplomat, physician, counselor — Copernicus often held several jobs at once. He was occasionally called on for special duties, such as his work on reforming the currency system to address the debasement of Prussian coinage. Many of these assignments were completed under the ever-present threat of war. Copernicus' life and work was often disrupted by the tasks of assisting in the planning of defenses for Frauenburg and surrounding areas during the ongoing wars with the Teutonic knights. Although his responsibilities as canon were not always time-consuming, other duties limited the time Copernicus had to make observations and work on his new theories.

An aging Copernicus continued to delay publication of his final work, in spite of encouragement, even pleas, from friends throughout Europe. The arrival of a young protégé in Frauenburg finally provided the impetus required to see the project through. Georg Joachim von Lauchen Rheticus, an Austrian mathematician and astronomer at the University of Wittenberg, left his university position to travel and study with the leading astronomers in Europe. In 1539, his travels took him to the door of Copernicus, where Rheticus remained for two years. Rheticus had been with Copernicus for only a short time when he procured from the mayor of Danzig financial assistance to help with the publication of the *Narratio Prima* (*First Account*), an overview of Copernicus' heliocentric theory.

Among many arguments made in favor of the Copernican model, Rheticus appealed to the relative simplicity of heliocentrism:

> Mathematicians as well as physicians, must agree with the statements emphasized by Galen here and there: "Nature does nothing without purpose" and "So wise is our Maker that each of his works has not one use, but two or three or often more." Since we see that this one motion of the earth satisfies an almost infinite number of appearances, should we not attribute to God, the creator of nature, that skill which we observe in the common makers of clocks? For they carefully avoid inserting in the mechanism any superfluous wheel or any whose function could be served better by another with a slight change of position.

The publication of the *First Account* helped spread the word of his teacher's work.

During his two-year stay in Poland, Rheticus encouraged Copernicus and helped him prepare his completed work for publication. In 1541, after currying favor with Duke Albert of Prussia with gifts of a map of Prussia and an instrument used to determine the length of the day, Rheticus received the approval of the duke to publish Copernicus' book. Before the final version appeared, however, Rheticus published a section containing some of the trigonometric calculations used by Copernicus, even adding trigonometric tables of his own making. Although Rheticus had already returned to another university position at Wittenberg, he took leave of this position in 1542 to travel to Nuremberg in order to personally supervise the printing of what would be Copernicus' masterpiece: *On the Revolutions of the Heavenly Spheres*. When his duties called him back to Wittenberg, Rheticus left the task of seeing the printing through to completion to the Lutheran theologian Andreas Osiander. Nicolaus Copernicus died on May 24, 1543: contemporary accounts tell us that he was given the first copy of his printed masterpiece as he lay on his deathbed.

The publication of *On the Revolutions of the Heavenly Spheres* in 1543 is often considered the first shot fired in the scientific revolution. Although its immediate impact was slight, its ideas, especially its central idea of a heliocentric cosmos, influenced generations of scientists who carried out this revolution in scientific thinking.

Key Aspects of *On the Revolutions*

First and foremost, *On the Revolutions* removed the earth from its privileged place at the center of the universe and replaced it with the sun. Copernicus wrote that the centrality and immobility of the sun represented its divine nature.

> At rest, however, in the middle of everything is the sun. For in this most beautiful temple, who would place this lamp in another or better position than that from which it can light up the whole thing at the same time? For, the sun is not inappropriately called by some people the lantern of the universe, its mind by others, and its ruler by still others.... Thus indeed, as though seated on a royal throne, the sun governs the family of planets revolving around it.

For Copernicus, it made sense that the sun occupied an honored place in God's creation.

He also calculated that the sun was only *near* the center of the universe: the actual center was the center of the earth's orbit, thus reserving, in a way, a special position for the earth in the cosmos. Copernicus attributed to the earth three movements: a diurnal (daily) motion spinning on its axis in order to explain the apparent daily movements of the sun, planets, and stars; a yearly revolution around the sun to explain movement on the ecliptic; and a third motion, which was an (ultimately unsuccessful) attempt by Copernicus to account for the changes in seasons and the precession of the equinoxes.

All of these motions attributed to the earth and other planets were conceived by Copernicus for one overriding purpose: to maintain the idea of perfect circular motion in the heavens. However, Copernicus also found that he must provide for motion by epicycles in order to "save the appearances." In other words, for his model to fit the observed motions — or at least very nearly fit the observed motions — epicycles were required. Allowing for epicycles meant the sacred circular motion could be preserved.

Copernicus realized that his heliocentric model implied that the universe was much more immense than previously conceived. The reason for this implication was the lack of observable stellar parallax. Parallax means an apparent motion of an object due to the actual motion of the observer. For example, hold your index finger at arm's length in front of your face and observe an object behind it at some distance. Now, close your left eye and observe the object with your right eye only. Next, close your right eye and observe the same object with your left eye only. The object appears to move in relation to your extended index finger, when in reality the position from which you observed the object was what moved. This is parallax. Stellar parallax means that the position of any particular star, as observed against a fixed background, should appear to change position as the earth makes its trek around the sun. Copernicus knew that this stellar parallax had never been observed, and concluded that the only explanation was that the distances to the stars were such that the apparent motion was undetectable from earth. As it turns out, stellar parallax was not detected until the development of powerful telescopes in the nineteenth century (the angle of parallax to the nearest star, it turns out, is less than $\frac{1}{5000}$ of a degree — a truly minuscule angle). It also turns out that even Copernicus was much too conservative in his estimation of the size of the universe.

Copernicus was motivated by several factors in publishing his findings in *On the Revolutions*. From a strictly physical point of view, he realized his heliocentric system explained certain observed phenomena much better than the Ptolemaic system. For instance, a heliocentric system explained the observed retrograde motion of the planets without resorting to ad hoc constructions like epicycles. Observers of the night sky had long been intrigued and baffled by retrograde motion. Each planet appeared to move eastward through the sky most of the time; however, at times these same planets appeared to reverse their direction and move to the west. This is retrograde motion. Ptolemy accounted for retrograde motion by putting planets on epicycles that moved the planets in the opposite direction of their natural motion. Copernicus realized that if the earth was moving along with the other planets, the changing position of the planets was natural whenever the earth passed an outer planet or was passed by an inner planet.

Copernicus' heliocentric model also settled a long-standing confusion concerning the relative position of Mercury and Venus. In the Ptolemaic system, the planetary order of these two inner planets was ambiguous. Was Mercury closest to the sun, or was Venus? Were these two planets between the earth and the sun, or on the other side of the sun? The Ptolemaic system did not answer these questions, causing many to feel uncomfortable with any system that allowed such ambiguity. The heliocentric model settled the question — Mercury was the first planet from the sun and Venus was the second. Copernicus had once again found an elegant solution to a messy problem.

In fact, a messy problem is just what Copernicus saw when he studied the Ptolemaic version of the cosmos. He wrote about this in the dedication of *On the Revolutions*:

> I was impelled to consider a different system of deducing the motions of the universe's spheres for no other reason than the realization that astronomers do not agree among themselves in their investigations of this subject ... their experience was just like some one taking from various places hands, feet, a head, and other pieces, very well depicted, it may be, but not for the representation of a single person; since these fragments would not belong to one another at all, a monster rather than a man would be put together from them.

Copernicus was obviously concerned with the aesthetics of the hodge-podge system developed over time. He described, almost three centuries before Mary Shelley, a cosmological Frankenstein's monster. He wrote his life work in the hopes of creating a simpler, more aesthetically pleasing model of the cosmos. To accomplish this, his first priority was to restore perfect circular motion to the paths of the heavenly bodies.

Before he proceeded with his arguments for circular motion, Copernicus first sought to establish that the universe, and all of the bodies contained therein, was spherical. We can see the mathematical, physical, and metaphysical underpinnings of Copernicus' cosmology in his argument for a spherical universe. This argument, presented in the first chapter of *On the Revolutions*, states:

First of all, we must note that the universe is spherical. The reason is either that, of all forms, the sphere is the most perfect, needing no joint and being a complete whole, which can be neither increased nor diminished; or that it is the most capacious of figures, best suited to enclose and retain all things; or even that all the separate parts of the universe, I mean the sun, moon, planets and stars, are seen to be of this shape; or that wholes strive to be circumscribed by this boundary, as is apparent in drops of water and other fluid bodies when they seek to be self-contained. Hence no one will question the attribution of this form to the divine bodies.

Having established that the universe must be spherical, Copernicus used similar metaphysical and observational arguments to establish the spherical nature of the earth. Finally, with the question of a spherical earth and a spherical universe firmly established, Copernicus turned to the question of natural motion for spherical bodies:

> The motion of the heavenly bodies is circular, since the motion appropriate to a sphere is rotation in a circle. By this very act the sphere expresses its form as the simplest body, wherein neither beginning nor end can be found, nor can the one be distinguished from the other, while the sphere itself traverses the same points to return upon itself.

Not only metaphysical, but also physical considerations required circular motion. In fact, in the eyes of the Polish astronomer, circular motion was the only appropriate explanatory structure for the irregularities in planetary motion:

> We must acknowledge, nevertheless, that their motions are circular or compounded of several circles, because these nonuniformities recur regularly according to a constant law. This could not happen unless the motions were circular, since only the circle can bring back the past.

Circular motion was an ancient and revered idea in astronomy. The philosophy of circular motion extends at least as far into antiquity as the Pythagoreans, if not farther. Aristotle held circular motion to be the natural motion of heavenly bodies, much in the same way as rectilinear motion was the natural motion of terrestrial bodies. But it wasn't only in the heavens that perfect circles were observed; the spherical earth itself represented a manifestation of this perfect geometry. Contrary to the misconception that Columbus set sail to prove the earth was round, the concept of a spherical earth was not disputed by philosophers and the educated. The Greeks argued that observations such as the disappearance of a ship's deck long before its mast as it sailed away showed the earth's surface was curved. Moreover, every time a lunar eclipse occurred, the earth's spherical shadow was evident on the moon's surface. That the earth was spherical, and the orbits of the heavenly bodies circular, was not questioned.

As we have seen, Copernicus found several points of contention with the Ptolemaic cosmos, but the most disconcerting remained his conviction that the Ptolemaic system violated the doctrine of circular motion. Ptolemy realized that the earth was not at the exact center of the planet's orbits; he therefore created a deferent around which the center of a planet's epicycle rotated. Still not satisfied that the results matched observations, Ptolemy inserted another point, called an equant, on the side of the deferent center opposite and equidistant from the earth. Uniform motion, according to the Ptolemaic system, occurred only with respect to the equant, not with respect to the earth (or any other celestial body). The convoluted motion caused by the existence of the equant, motion that seemed blatantly non-circular, was abhorrent to Copernicus. He was determined to remedy this situation.

The Preface and Introduction to *On the Revolutions*

When *On the Revolutions of the Heavenly Spheres* was finally published in 1543, in the same year as Copernicus' death, it included a preface that was obviously meant to soften the blow of such a controversial subject. This preface said that *On the Revolutions* should be read as a mathematical theory, not an attempt to describe the physical reality of the world. Furthermore, it claimed that the astronomer's job was to create a predictive tool that coincided with observational data: "For these hypotheses need not be true nor even probable. On the contrary, if they provide a calculus consistent with the observations, that alone is enough." The preface went so far as to say that Copernicus' cosmology was no more or less certain than that of previous astronomers: "Therefore alongside the ancient hypotheses, which are no more probable, let us permit these new hypotheses also to become known.... So far as hypotheses are concerned, let no one expect anything certain from astronomy."

These words, found in the very first paragraphs of *On the Revolutions*, do not sound like the words of a man who had dedicated his life to developing a new approach to the structure of the heavens. That's because they weren't. When Rheticus left to take a position as professor of mathematics at the University of Leipzig, he turned over the publication of *On the Revolutions* to a Lutheran minister by the name of Andreas Osiander. It was Osiander who added the preface, without the knowledge and consent of Copernicus or of the two men who helped Copernicus through the publication process, Rheticus and Petreius. To readers of *On the Revolutions* who were not privy to Osiander's sleight of hand, it appeared that Copernicus had written the preface and in the process contradicted his own work. It was not until 1609 that Johannes Kepler publicly announced the identity of the author of this preface, in his *Astronomia Nova*.

Following Osiander's unauthorized preface, we find Copernicus' first words written for *On the Revolutions*, a dedication to "His Holiness, Pope Paul III." Such a dedication was not at all unusual for authors of the time period. It was important to seek the patronage of powerful people, and of course the pope was the most powerful man in the world. Church officials, including Pope Paul III, were also generally well-educated humanists interested in a wide range of philosophy, including the sciences. In addition, Copernicus felt it necessary to lay out for the pope — and anyone else who read his book — justification for such a theory and a preliminary explanation of just what he would be claiming as truth. In the process, Copernicus also revealed some of his own personal motivation for writing what he knew would be a controversial work. It seems that Copernicus ascribed to the modern sports maxim, a good defense is a strong offense; several instances of preemptive attacks on his own detractors are found in this section. We will explore a few of these themes in the dedication of *On the Revolutions*.

Very early in the dedication, Copernicus explained that he has delayed publication of *On the Revolutions* while he struggled with concerns of how it would be received:

> Those who know that the consensus of many centuries has sanctioned the conception that the earth remains at rest in the middle of the heaven as its center would, I reflected, regard it as an insane pronouncement if I made the opposite assertion that the earth moves. Therefore I debated with myself for a long time whether to publish the volume which I wrote to prove the earth's motion.

Copernicus compared his plight to "the Pythagoreans" who "wanted the very beautiful thoughts attained by great men of deep devotion not to be ridiculed by those who are reluctant to exert themselves vigorously." Like the Pythagoreans, "who used to transmit philosophy's secrets only to kinsmen and friends, not in writing but by word of mouth," Copernicus was reticent about publishing. In fact, Copernicus wrote, "the scorn which I had reason to fear on account of the novelty and unconventionality of my opinion almost induced me to abandon completely the work which I had undertaken." It is clear from these statements that Copernicus struggled for some time with the idea of publishing his work.

However, in order to justify his musings on an earth in motion, Copernicus appealed to the ancient philosophers much revered in early modern Europe. He wrote: "I undertook the task of rereading the works of all the philosophers which I could obtain to learn whether anyone had ever proposed other motions of the universe's spheres than those expounded by the teachers of astronomy in the schools." His studies, Copernicus continued, led him to several such ancient philosophers, including a certain Hicetas (mentioned in Cicero) and several others found in the writings of Plutarch: Philolaus the Pythagorean, Heraclides of Pontus, and Ecphantus the Pythagorean. Throughout the text of *On the Revolutions*, Copernicus continued to refer to ancient authority, especially the Pythagoreans, capitalizing on the humanist philosophy that recovery of ancient knowledge was at the heart of intellectual endeavors.

It wasn't only namedropping of ancient authorities Copernicus utilized to bolster his reliability; he also wrote of the support received from Nicholas Schönberg, the cardinal of Capua; Tiedemann Giese, bishop of Chelmno; and "not a few other very eminent scholars." All of these scholars, ancient and contemporary, lent credence to Copernicus and his radical idea.

Copernicus also briefly explained to Pope Paul III the evolution of his idea that the earth moved. Of major concern was the seemingly chaotic state of sixteenth-century astronomy. Copernicus wrote, "I was impelled to consider a different system of deducing the motions of the universe's spheres for no other reason than the realization that astronomers do not agree among themselves in their investigations of this subject." Copernicus delineated these difficulties. First, astronomers were unable to accomplish the most basic tasks in accurate calendar making: "they are so uncertain about the motion of the sun and moon that they cannot establish and observe a constant length even for the tropical year." Secondly, astronomers "do not use the same principles, assumptions, and explanations of the apparent revolutions and motions." Copernicus was especially disturbed by the use of various techniques that "apparently contradict the first principles of uniform motion."

These disagreements among astronomers led to the monstrous system that so concerned Copernicus. He thus had justification for a new approach to the problem:

> And even though the idea seemed absurd, nevertheless I knew that others before me had been granted the freedom to imagine any circles whatever for the purpose of explaining the heavenly phenomena. Hence I thought that I too would be readily permitted to ascertain whether explanations sounder than those of my predecessors could be found for the revolution of the celestial spheres on the assumption of some motion of the earth.

His solution to the problem?

Having thus assumed the motions which I ascribe to the earth later on in the volume, by long and intense study I finally found that if the motions of the other planets are correlated with the orbiting of the earth, and are computed for the revolution of each planet, not only do their phenomena follow therefrom but also the order and size of all the planets and spheres, and heaven itself is so linked together that in no portion of it can anything be shifted without disrupting the remaining parts and the universe as a whole.

Instead of a monstrous system, Copernicus found a harmonious, aesthetically pleasing, and mathematically elegant cosmological theory.

One of the most interesting aspects of the dedication of *On the Revolutions* is the method by which Copernicus attacks his would-be detractors. He warned Pope Paul that there will be those who reject his work, "yet because of their dullness of mind they play the same part among philosophers as drones among bees." In addition, Copernicus warns that "there will be babblers who claim to be judges of astronomy although completely ignorant of the subject and, badly distorting some passage of Scripture to their purpose, will dare to find fault with my undertaking and censure it."

Dull, drones, babblers, ignorant — these are the words used by Copernicus to describe his enemies. On the other hand, those who agree with his work are described in a very different manner. "I have no doubt that acute and learned astronomers will agree with me if, as this discipline especially requires, they are willing to examine and consider, not superficially but thoroughly, what I adduce in this volume in proof of these matters."

Copernicus summarized this line of thought with an enduring maxim: "Astronomy is written for astronomers." His work, Copernicus was saying, was not intended for the mathematically ignorant (a category in which Copernicus places even the well-educated university professors and church officials who disagree with him); rather, *On the Revolutions* was written for a few experts able to comprehend its significance.

At this point it should be emphasized that *On the Revolutions* was, and is, an extremely difficult, technical, mathematical work. Although our current objective is to create a non-technical overview of Copernicus' work, it would be a great disservice to readers to leave the impression that this revolutionary book was anything but difficult to comprehend. At the risk of frightening away the general readership, I include at this point a single paragraph from *On the Revolutions*—without comment or explanation—to drive home the point that this is a work of mathematical astronomy:

However, as an example I shall use Mars, because it exceeds all the other planets in latitude. Its maximum southern latitude was noted by Ptolemy as about 7° when Mars was at perigee, and its maximum northern latitude at apogee as 4° 20' [*Syntaxis*, XIII, 5]. However, having determined angle *BGD* = 6° 50', found the corresponding angle *AFC* ~ 4° 30'. Given *EG* : *ED* = 1p : 1p 22' 26" [V, 19], from these sides and angle *BGD* we shall obtain angle *DEG* of the maximum Southern inclination ~ 1° 51'. Since *EF* : *CE* = 19 : 1° 39' 57" [V, 19] and angle *CEF* = *DEG* = 1° 51', consequently the aforementioned exterior *CFA* = 4½° when the planet is in opposition.

Copernicus' admonition that his work was meant for mathematicians rings as true today as it did over four hundred years ago.

Copernicus closed his dedication with a short note concerning a practical application of his work. Calendar reform had been foremost in the mind of many sixteenth-century astronomers and church officials. In fact, a council summoned by Pope Julius II sought the help of Copernicus in 1514 to revise the Julian calendar. Primarily motivated by a continuing problem with the proper placement of Easter, resolution only occurred with the adoption of the Gregorian calendar in 1582, almost four decades after the publication of *On the Revolutions*. Although Copernicus remained hopeful that his work would contribute to calendar reform, he promised nothing, writing, "what I have accomplished in this regard, I leave to the judgement of Your Holiness in particular and of all other learned astronomers."

Reactions to *On the Revolutions*

The immediate reaction to Copernicus's work was surprisingly mild. First, it must be remembered that this was not a bomb dropped on the learned philosophers or the church: three decades earlier Copernicus had circulated the *Commentariolus* among European philosophers and astronomers, and his disciple — Rheticus — published *Narratio Prima* in 1540 in which he outlined his teacher's theory. Therefore, there was no shock effect from a revolutionary idea no one had heard about previously. Secondly, following the foreword written by Osiander, many accepted Copernicus' work for its computational value without concern for the physical (or metaphysical) implications of a heliocentric system. Finally, the immediate response of the Catholic Church was positive. Copernicus had the support, even encouragement, of several high-ranking church officials including Cardinal Nicholas Schönberg. Although eventually placed on the list of banned works, *On the Revolutions* elicited no such response for the first few decades after its publication.

Arguments against a moving earth, however, were plentiful. The first arguments were those that appealed to common sense and everyday observation. Does the sun move? Of course it does! When I awoke this morning, it was on the eastern horizon. When I ate my noonday meal, it was directly overhead. Nightfall arrived as the sun set on the western horizon. I *saw* it move throughout the day. Well then, what about the earth — does it move? Of course it doesn't. If the earth moved, would we not feel it moving? When I ride a fast horse, I feel the wind in my face. If the earth spun on its axis, there would be a prevailing wind: birds would struggle to fly. If the earth moved, would not a stone thrown straight into the air land some distance behind as the earth spun beneath it? These are not arguments that require a great philosopher; any uneducated peasant can see that Earth is stationary and the moon, sun, planets, and stars revolve around it.

Yet there were also arguments of a more philosophical (scientific) bent offered by the university professors. One was the absence of stellar parallax, as mentioned above. Opponents of the heliocentric theory argued that the lack of an observable stellar parallax meant that Copernicus' theory was not true. Copernicus, and Copernicans, maintained that this indicated the immense size of the universe: stellar parallax did exist but the great distances to the stars meant that the angles were much too small to observe. Another argument made by the astronomers was that if the earth moved around the sun, with Venus also orbiting the sun inside Earth's orbit, Venus should exhibit phases exactly like the moon. In other words, we should be able to observe a half-Venus, a quarter-Venus, etc., just as we see a half-moon and a quarter-moon. Since Venus did not exhibit these phases, Copernicus must be wrong. (Not until Galileo turned his telescope to Venus was this argument answered.)

For the learned natural philosophers of the European universities, the most damning evidence against Copernicus was the fact that it contradicted the very basis of Aristotle's physics. Recall that in Aristotelian physics, heavy things fell to the earth because they sought their natural place. More importantly, heavy things are heavy because they are made primarily of the heaviest element: earth. The heavy earth is immobile and central to Aristotelian physics just as it is immobile and central to Aristotelian (and Ptolemaic) astronomy. If the earth moves, the very premise of physics is violated, and all of natural philosophy (physics) is wrong. Now, to propose that a cherished theory is incorrect is one thing, but to have no theory ready to offer in its stead is another thing altogether. Copernicus offered no replacement for the system of natural philosophy that had been accepted by the academic community for centuries! Copernicus had proposed a system that would overthrow the most cherished precepts of traditional science, and had not offered a valid replacement. (Once again, Galileo later played an important role in establishing a new physics that did not conflict with the Copernican cosmology.)

Finally, the heliocentric theory of Copernicus was in opposition to church traditions and teachings, and even to the Bible itself:

> Then spake Joshua to the LORD in the day when the LORD delivered up the Amorites before the children of Israel, and he said in the sight of Israel, sun, stand thou still upon Gibeon; and thou, moon, in the valley of Ajalon. And the sun stood still, and the moon stayed, until the people had avenged themselves upon their enemies. Is not this written in the book of Jasher? So the sun stood still in the midst of heaven, and hasted not to go down about a whole day.*

When Joshua needed extended daylight to complete his victory, did he ask God to make the earth stop spinning? No — he asked God to cause the sun to stop moving. Proof enough that the system contrived by Copernicus was not the system devised by the creator of the universe.

Even more than the single passage in Joshua, a geocentric cosmos fit the notion that man was the center of God's creation. As the culmination of the Genesis creation story, and as the figurative center of God's plan for the universe, should not man also occupy the literal center of the universe, the earth?

Copernicans, both contemporaries of and subsequent to Copernicus himself, countered these arguments against the heliocentric theory with arguments of their own. Why, you ask, do we not feel the movement of the earth? Why is there no sense of motion, no prevailing winds? The answer is simple: motion is relative and the atmosphere of the earth is moving with the earth itself. Copernicus, anticipating such arguments, wrote in *On the Revolutions*, "Every observed change of place is caused by motion of either the observed object or the observer or, of course, by an unequal displacement of each." Furthermore, Copernicus contended that "if any motion is ascribed to the earth, in all things outside it the same motion will appear, but in the opposite direction, as though they were

*Joshua 10:12–13. This is from the King James Version of the Bible, first published in 1611.

moving past it." The concept of relative motion — the motion of the observer in relation to the observed — answered the critics of the moving earth theory.

The Copernican system also fit observations in a way that the Ptolemaic system did not. As we have already seen, placing the earth in motion around the sun provides a ready-made — and elegantly simple — explanation for retrograde motion and for the order and relative distances of Mercury and Venus.

Copernicans also turned around the argument that a heavy earth should not move. They said that the earth is but a small speck in the vast universe. Copernicus once again anticipated this difficulty: "the heavens are immense by comparison with the earth and present the aspect of an infinite magnitude, while ... the earth is related to the heavens as a point to a body, and a finite to an infinite magnitude." Why should this tiny body remain motionless while the vast universe spins around at dizzying speeds? Would it not make more sense to say the sphere of the stars remains motionless (along with the sun) while the earth moves? "Indeed, a rotation in twenty-four hours of the enormously vast universe should astonish us even more that a rotation of its least part, which is the earth." Later, Copernicus returned to an argument concerning relative motion, noting:

> For when a ship is floating calmly along, the sailors see its motion mirrored in everything outside, while on the other hand they suppose that they are stationary, together with everything on board. In the same way, the motion of the earth can unquestionably produce the impression that the entire universe is rotating.

Each of the points made by the Copernicans supported one of Copernicus' underlying motivations, to provide for a more aesthetic system than the monstrous Ptolemaic system.

The Copernican Revolution?

In 1957, Thomas Kuhn published a seminal work in the history of science, *The Copernican Revolution*: *Planetary Astronomy in the Development of Western Thought*. Kuhn later (1962) published perhaps the most influential, and most debated, work in the history of science, *The Structure of Scientific Revolutions*. Since then, historians have questioned whether a scientific revolution occurred, as opposed to a theory that argues for more continuous evolution in scientific thought. That question aside, we might ask, "Was Copernicus himself a revolutionary figure?" Many historians think not. Paulo Rossi, in his book *The Birth of Modern Science*, wrote, "Copernicus never took a revolutionary stance either personally or professionally."

What of Copernicus himself? Was he self-consciously creating a revolution in science? We may find hints of his intentions in the frontpiece of *On the Revolutions*, in which Copernicus wrote of both the theoretical and practical nature of his masterpiece:

> Diligent reader, in this work, which has just been created and published, you have the motions of the fixed stars and planets, as these motions have been reconstituted on the basis of ancient as well as recent observations, and have moreover been embellished by new and marvelous hypotheses. You will also have most convenient tables, from which you will be able to compute those motions with the utmost ease for any time whatever. Therefore, buy, read, and enjoy [this work].

As further proof of the practical, mathematical nature of his work, Copernicus added the byline, "Let no one untrained in geometry enter here." This line, traditionally believed to be a sign over the entrance to Plato's Academy, warned readers of the technical difficulty of the work but certainly makes no claims as to the revolutionary nature of his hypotheses.

There are many reasons for claiming that Copernicus was not revolutionary. Copernicus was a dedicated humanist — in a culture dominated by humanism — looking for answers in ancient texts. Rather than striving to create new ideas and new knowledge, Copernicus was attempting to revive the work of the ancient authorities, especially the Pythagoreans. In addition, Copernicus looked upon Ptolemy as a guide in his work. He wrote *On the Revolutions* using the same structure as the *Almagest*. In fact, Johannes Kepler later claimed that Copernicus interpreted Ptolemy rather than nature. Very little about the Polish canon infers revolutionary; yet, *On the Revolutions* opened new horizons that culminated in a revolutionary way of understanding our natural world.

PRIMARY SOURCES

Copernicus, Nicholas. *On the Revolutions of the Heavenly Spheres*, translated by Edward Rosen. London: Macmillan, 1972.

_____. *Three Copernican Treatises: The Commentariolus of Copernicus, the Letter Against Werner, the Narratio Prima of Rheticus*, translated by Edward Rosen. London: Octagon Books, 1971.

OTHER SOURCES

Bienkowska, B. *The Scientific World of Copernicus on the Occasion of the 500th Anniversary of His Birth 1473–1973*. Dordrecht and Boston: D. Reidel, 1972.

Danielson, Dennis Richard. *The First Copernican: Georg Joachim Rheticus and the Rise of the Copernican Revolution*. New York: Walker, 2006.

Gingrich, Owen. *The Book Nobody Read: Chasing the Revolutions of Nicolaus Copernicus*. New York: Walker, 2004.

Kuhn, Thomas. *The Copernican Revolution: Planetary Astronomy in the Development of Western Thought*. Cambridge: Harvard University Press, 1957.

_____. *The Structure of Scientific Revolutions*. Chicago: University of Chicago Press, 1962.

Margolis, Howard. *It Started with Copernicus: How Turning the World Inside Out Led to the Scientific Revolution*. New York: McGraw-Hill, 2002.

Moss, Jean Dietz. *Novelties in the Heavens: Rhetoric and Science in the Copernican Controversy*. Chicago: University of Chicago Press, 1993.

Repcheck, Jack. *Copernicus' Secret: How the Scientific Revolution Began*. New York: Simon & Schuster, 2007.

Rosen, Edward. *Copernicus and the Scientific Revolution*. Malabar, FL: Krieger, 1984.

Vollmmann, William T. *Uncentering the Earth: Copernicus and the Revolutions of the Heavenly Spheres*. New York: Norton, 2006.

Westfall, Robert S. *The Copernican Achievement*. Berkeley: University of California Press, 1975.

CHAPTER 2

Galileo: Astronomy, the Birth of Modern Physics, and Science's Battle with the Church

That the intention of the Holy Ghost is to teach us how one goes to heaven, not how heaven goes.—Galileo Galilei, letter to Christina, grand duchess of Tuscany (1615)

Copernicus published *On the Revolutions of the Heavenly Spheres* in 1543, yet the turn of the century came and went without much evidence to support his revolutionary thesis. The heliocentric system did gain its converts for its simplicity and mathematical utility, but even the church seemed to think the theory was innocuous, failing to ban the text until 1616. By this time, new and controversial evidence surfaced supporting the Copernican system and began a battle between the conservatives and traditionalists and the adherents to the new science.

Much of this new evidence was thanks to the shocking discoveries of the Italian mathematician Galileo Galilei. Not only did Galileo reveal a new universe under the scrutiny of the recently invented telescope, he also began a revolution in natural philosophy by making physics an experimental science. Galileo's works ushered in a new age for cosmology and for the study of science, paving the way for men such as Isaac Newton to further unlock the secrets of the universe.

Early Life of Galileo

Galileo was born in Pisa in 1564. His father, Vincenzo Galilei, was an accomplished musician who applied his knowledge of mathematics to the study of music. Galileo inherited his father's musical talent, as well as a lifelong interest in art and poetry. As a young man, Galileo entertained the notion of entering the priesthood but was discouraged from this by his father who instead sent him to the University of Pisa to obtain a medical degree. Young Galileo, however, had other plans. His interest in mathematics eventually led him to drop his study of medicine, and with the help of his teacher, Ostilio Ricci, Galileo eventually convinced his father to let him study mathematics. Although he never completed his degree, he was soon tutoring students in mathematics and in 1589 was appointed to the chair of mathematics at Pisa. This was not a well paying position. In 1591 the young mathematician found himself in need of more income when his father died, leaving Galileo as the head of the family and responsible for the dowries of his younger sisters. So in 1592 Galileo moved to the University of Padua where his income was significantly increased, although still considerably less than that of a professor of philosophy and much less than the self-confident Galileo believed he deserved. The search for a higher income through invention and patronage became a life-long pursuit for Galileo, a need sharpened by the eventual birth of three illegitimate children Galileo fathered with Marina Gamba.* Galileo stayed at Padua until 1610, and it was here that he began making a name for himself in the scientific community.

Galileo: Astronomer

Just who invented the telescope is not clear. In the first decade of the seventeenth century several glassmakers in the Netherlands laid claim to the invention. Although Galileo never claimed to have originated the idea, he is often presented as the inventor of the telescope. This misconception might come from the title page of *The Starry Messenger*, a book published by Galileo in 1610 detailing his magnificent discoveries made with the telescope:

> THE STARRY MESSENGER, Revealing great, unusual, and remarkable spectacles, opening these to the consideration of every man, and especially of philosophers and astronomers; AS OBSERVED BY GALILEO GALILEI. Gentleman of Florence, Professor of Mathematics in the University of Padua, WITH THE AID OF A SPYGLASS *lately invented by him,* In the surface of the Moon, in innumerable Fixed Stars, in Nebulae, and above all in FOUR PLANETS swiftly revolving about Jupiter at differing distances and periods, and known to no one before the Author recently perceived them and decided that they should be named THE MEDICEAN STARS [Venice 1610].

The phrase, "lately invented by him," most certainly refers to the fact that Galileo had labored to improve the original device he had received and succeeded in making a better, more powerful telescope, or spyglass:

> About ten months ago a report reached my ears that a certain Fleming had constructed a spyglass by means of which visible objects, though very distant from the eye of the observer, were distinctly seen as if nearby. Of this truly remarkable effect several experiences were related, to which some persons gave credence while others denied them. A few days later the report was confirmed to me in a letter from a noble Frenchman at Paris, Jacques Badovere, which

---

*One of these children, a daughter christened Virginia, provided historians with a treasure trove of insight into Galileo's life and works. Placed in a convent at the age of 13 (a fate shared with her sister as Galileo felt that this was the best way to protect his illegitimate daughters), Virginia took the name Marie Celeste and began a life-long correspondence with her father in which Galileo shared not only his work, but also the most intimate details of his life. See Dana Sobel's *Galileo's Daughter.*

caused me to apply myself wholeheartedly to inquire into the means by which I might arrive at the invention of a similar instrument. This I did shortly afterwards, my basis being the theory of refraction. First I prepared a tube of lead, at the ends of which I fitted two glass lenses, both plane on one side while on the other side one was spherically convex and the other concave. Then placing my eye near the concave lens I perceived objects satisfactorily large and near, for they appeared three times closer and nine times larger than when seen with the naked eye alone. Next I constructed another one, more accurate, which represented objects as enlarged more than sixty times. Finally, sparing neither labor nor expense, I succeeded in constructing for myself so excellent an instrument that objects seen by means of it appeared nearly one thousand times larger and over thirty times closer than when regarded with our natural vision.

Galileo's first thoughts as to the utility of his device were not directed at the heavens. Rather, Galileo had a much more immediately useful, and potentially lucrative, plan for the telescope. Possessing an instrument that could detect a ship at sea long before it appeared to the naked eye had obvious benefits to a seagoing people such as the Venetians. So Galileo, ever mindful of his monetary needs, demonstrated his new telescope to the Venetian Senate and presented the device to them in return for a large increase in his salary at Padua. This was not the first time Galileo had attempted to profit from his mechanical skills. Previously, he had invented (and profited from) a military compass that proved a great help in improving the accuracy of cannon.

Soon, Galileo turned his newly constructed telescope to the heavens — an event that changed his life and changed history. With the aid of the telescope, Galileo witnessed things in the heavens never before seen by earthbound humans and in the process provided startling new evidence for the Copernican heliocentric theory. In the end, the ideas Galileo gleaned from his observations led not only to his everlasting place in history, but more immediately to a conflict with the church that ended badly for Galileo.

One of the first heavenly objects upon which Galileo turned his telescope was, for obvious reasons, the moon. What he observed was astonishing. Aristotelian doctrine held that the moon was in the uncorrupted region above the realm of the earth. This doctrine maintained that the moon was not made of the same elements as the earth, but was perfect, smooth, and unchanging — very unearthlike. On the contrary, Galileo wrote in *The Starry Messenger*: "The moon is not robed in a smooth and polished surface but is in fact rough and uneven, covered everywhere, just like the earth's surface, with huge prominences, deep valleys, and chasms."

Galileo continued later in his book:

I have been led to the opinion and conviction that the surface of the moon is not smooth, uniform, and precisely spherical as a great number of philosophers believe it (and the other heavenly bodies) to be, but is uneven, rough, and full of cavities and prominences, being not unlike the face of the earth, relieved by chains of mountains and deep valleys.

Galileo confirmed with his telescope what a few had dared to postulate based on naked eye observations — the moon appears to be just like the earth.

Galileo also turned his telescope to the Milky Way, that great cloud-like (or milky) apparition in the night sky. Again, under magnification the results were astounding. Instead of a cloudy, undefined glob of material, Galileo discovered that the Milky Way was actually composed of untold numbers of individual stars. He recognized what it meant to make such a discovery:

Surely it is a great thing to increase the numerous host of fixed stars previously visible to the unaided vision, adding countless more which have never before been seen, exposing these plainly to the eye in numbers ten times exceeding the old and familiar stars. Again, it seems to me a matter of no small importance to have ended the dispute about the Milky Way by making its nature manifest to the very senses as well as to the intellect.

But Galileo wasn't finished. The entire universe seemed to open up much like the Milky Way under the intense scrutiny of Galileo and his telescope. Everywhere he looked he discovered more stars:

The galaxy is, in fact, nothing but a congeries of innumerable stars grouped together in clusters. Upon whatever part of it the telescope is directed, a vast crowd of stars is immediately presented to view. Many of them are rather large and quite bright, while the number of smaller ones is quite beyond calculation. But it is not only in the Milky Way that whitish clouds are seen; several patches of similar aspect shine with faint light here and there throughout the aether, and if the telescope is turned upon any of these it confronts us with a tight mass of stars. And what is even more remarkable, the stars which have been called "nebulous" by every astronomer up to this time turn out to be groups of very small stars arranged in a wonderful manner.

Not only had Galileo clearly shown the true nature of the Milky Way, he found evidence that the universe was much larger and contained many more stars than anyone had ever imagined. And these discoveries were only the beginning of the unimaginable sights first viewed by the Italian mathematician.

Greek astronomers differentiated between two types of stars. The fixed stars, according to Ptolemy and accepted by Europeans for centuries, revolved around the earth on a fixed sphere on the outermost shell of the universe. The wandering stars, or planets (planet is from the Greek for wanderer), were those stars that did not remain fixed relative to the cosmic background. These planets were seven in all — the sun, the moon, Mercury, Venus, Mars, Jupiter, and Saturn — each of which was observable to the naked eye. When Galileo trained his telescope first on the fixed stars and then on the planets, he discovered an astonishing difference:

The planets show their globes perfectly round and definitely bounded, looking like little moons, spherical and flooded all over with light; the fixed stars are never seen to be bounded by a circular periphery, but have rather the aspect of blazes whose rays vibrate about them and scintillate a great deal.

In other words, the planets came into sharp focus when viewed through the telescope while the fixed stars did not — they continued to twinkle. This observation had a significant impact on one of the most common arguments against the Copernican system. As discussed in Chapter 1, if the earth did revolve around the sun, the relative position of the fixed stars should change as the earth moved from one spot in its orbit to another — a phenomenon called stellar parallax. Since such a movement was not observed, it must imply, argued the Ptolemaic traditionalists, that the earth did not move. However, one implication of Galileo's observation that the planets came into focus under the magnification of his telescope while the fixed stars did not was that the stars were *much* farther away than the planets, not simply on a sphere some relatively short distance past the sphere of the last known planet, Saturn. This would explain the lack of stellar parallax — the stars are simply much too far away for such a movement to be detected.

For Galileo, the most exciting discovery he made with his telescope was that Jupiter had moons just like the earth had a moon. Galileo described his discovery in *The Starry Messenger*:

> But what surpasses all wonders by far, and what particularly moves us to seek the attention of all astronomers and philosophers, is the discovery of four wandering stars not known or observed by any man before us. Like Venus and Mercury, which have their own periods about the sun, these have theirs about a certain star that is conspicuous among those already known, which they sometimes precede and sometimes follow, without ever departing from it beyond certain limits.

Later, Galileo reiterates his perception of the importance of the discovery of the moons of Jupiter:

> There remains the matter which in my opinion deserves to be considered the most important of all the disclosure of four PLANETS never seen from the creation of the world up to our own time, together with the occasion of my having discovered and studied them, their arrangements, and the observations made of their movements and alterations during the past two months.

It is significant that Galileo compares the orbits of the moons around Jupiter to the orbits of the planets Venus and Mercury around the sun. The theory, almost universally accepted since the Greeks, was that the universe had one center of rotation — the earth. That all heavenly bodies revolved around the earth formed the basis of the cosmology, astronomy, and even the physics of the natural philosophers of Galileo's time. If it were true that some bodies revolved around Jupiter and not earth, it opened up to question some ancient — and dearly held — assumptions.

Before going into some detail concerning his discovery of Jupiter's moons, Galileo issued a challenge to other astronomers:

> I invite all astronomers to apply themselves to examine them and determine their periodic times, something which has so far been quite impossible to complete, owing to the shortness of the time. Once more, however, warning is given that it will be necessary to have a very accurate telescope such as we have described at the beginning of this discourse.

Because Galileo was in a hurry to publish his findings, he did not take the time to accurately calculate the periods of the moons' orbits. What followed, however, was a fascinating account of Galileo's discovery of these moons. The reader of *The Starry Messenger* feels like he is sneaking a look into Galileo's diary, as he records his daily observations and thoughts concerning what he sees:

> On the seventh day of January in this present year 1610, at the first hour of night, when I was viewing the heavenly bodies with a telescope, Jupiter presented itself to me; and because I had prepared a very excellent instrument for myself I perceived (as I had not before, on account of the weakness of my previous instrument) that beside the planet there were three starlets, small indeed, but very bright. Though I believed them to be among the host of fixed stars, they aroused my curiosity somewhat by appearing to lie in an exact straight line parallel to the ecliptic, and by their being more splendid than others of their size.

From Galileo's entry for January 7, 1610, it is obvious that at first he thought he had simply discovered three new stars in the same vicinity of the sky occupied by Jupiter.

Remember, Galileo had discovered an abundance of new stars, and since he had no way of calculating the distance of the stars he assumed these bodies were not actually close to Jupiter but instead a far distance in the sky among the other fixed stars. However, something did catch his attention and he was anxious to come back the next night to study the new stars. His curiosity piqued, Galileo returned his telescope to Jupiter the following night, only to find the stars in a different position relative to Jupiter. Now Galileo was becoming even more curious, if not confused, writing, "Hence it was with great interest that I awaited the next night."

We can sense Galileo's excitement, and imagine he anxiously awaited nightfall to get back to his observations. And what did he see? "But I was disappointed in my hopes, for the sky was then covered with clouds everywhere." Of course, the sky must eventually clear and when it did on the following evening, Galileo was back at his telescope pointing it towards Jupiter. Galileo's observations of the moons of Jupiter continued almost nightly through the first days of March. He discovered one more star in the vicinity of Jupiter, and eventually came to the realization that he was not actually looking at stars, but rather at moons orbiting Jupiter:

> I had now decided beyond all question that there existed in the heavens three stars wandering about Jupiter as do Venus and Mercury about the sun, and this became plainer than daylight from observations on similar occasions which followed. Nor were there just three such stars; four wanderers complete their revolutions about Jupiter, and of their alterations as observed more precisely later on we shall give a description here.

Galileo, of course, immediately realized the implications of his discovery. The discovery would be a rebuttal to one of the arguments made against the Copernican system. Critics maintained that a system that required the moon to revolve around the earth while at the same time the earth was orbiting the sun was nonsensical and physically impossible. Galileo argued:

> Here we have a fine and elegant argument for quieting the doubts of those who, while accepting with tranquil mind the revolutions of the planets about the sun in the Copernican system, are mightily disturbed to have the moon alone revolve about the earth and accompany it in an annual rotation about the sun. Some have believed that this structure of the universe should be rejected as impossible. But now we have not just one planet rotating about another while both run through a great orbit around the sun; our own eyes show us four stars which wander around Jupiter as does the moon around the earth, while all together trace out a grand revolution about the sun in the space of twelve years.

Ever the struggling scientist in search of powerful patrons, Galileo named his recently discovered celestial bodies the Medicean stars, after the powerful ruling family of Tuscany. As a boy, Cosimo (II) dé Medici was tutored by Galileo and continued to hold the mathematician in high esteem. Fortunately for Galileo, Cosimo (II) was now the grand duke of Tuscany and, conveniently, had three brothers — four dé Medicis for four newly discovered satellites of Jupiter! Did the ploy work? Barely one month after the publication of *The Starry Messenger*, Galileo was offered the position of chief mathematician at the University of Pisa and, more importantly, chief philosopher and mathematician to the grand duke of Tuscany. Such a position carried with it much prestige, a generous salary, and very few duties — leaving Galileo the leisure time to pursue his studies.

The publication of *The Starry Messenger* made a tremendous impression in Italy and indeed throughout Europe. Although soon Galileo's support of the Copernican system came under attack by the Inquisition, initially Galileo was received as somewhat of a celebrity. During a trip to Rome to demonstrate his telescope (and his discoveries made with the instrument) to Jesuit scholars, Galileo became a member of the Lyncean Academy, one of the earliest scientific societies in Europe.

Galileo's work with the telescope did not end with the publication of *The Starry Messenger*. After many more months of careful observation and calculation, Galileo accomplished what he had earlier challenged other astronomers to do: he accurately calculated the orbits of the moons of Jupiter. Galileo went on to make other very significant discoveries with the telescope. One of these discoveries arose as he studied Venus.

One of the implications of the heliocentric model of the universe was that Venus should exhibit phases much like the moon because of the relative positions of the Sun, Venus, and Earth. In the Ptolemaic system, on the other hand, these phases were not predicted. Since Venus did not appear to go through phases, it seemed a pertinent piece of evidence in favor of tradition and Ptolemy. That is until Galileo began studying Venus through the telescope. When magnified, Galileo witnessed phases of Venus just as predicted by the heliocentric model. These observations of Venus, perhaps more than any of Galileo's other telescopic discoveries, resulted in more astronomers adopting the heliocentric model of the cosmos. In fact, this discovery seems to have fueled the fire of Galileo's own heliocentric leanings, as he became bolder in maintaining the truth of the Copernican theory.

Having already found exciting new phenomena while studying Jupiter and Venus, it made sense to next turn his telescope to another planet, in this case Saturn. What Galileo saw puzzled him: his telescope was not powerful enough to pick up the detail of Saturn's rings; instead, he thought he had detected two small planets at the edges of Saturn. To add to the confusion, when Saturn was positioned in a manner that the rings were aligned with Earth, the small planets disappeared. It took many years (and much more powerful telescopes) before the true nature of Saturn's rings was ascertained.

Galileo's priority in the discovery of these wondrous new properties of the cosmos can hardly be disputed. However, that is not the case with his discovery of sunspots. Although sunspots are detectable without a telescope and had been mentioned occasionally throughout history, it was not until Galileo and others turned the new instrument to the sun that careful and considered observations could be made. Not only did a dispute arise between Galileo and others over priority in the discovery of the sunspots, Galileo also became embroiled in a dispute concerning the nature of sunspots. If they were, as Galileo argued, actual spots on or near the sun's surface — moving and changing places from day to day — they would contradict the traditional Aristotelian view of the immutability of the heavens.

Galileo was convinced that all of the discoveries made with his telescope were evidence that Copernicus was correct and that the earth and the other planets orbited the sun. Interestingly, Galileo came to this conclusion himself long before his experiences with the telescope. In a 1597 letter to the German astronomer Johannes Kepler, Galileo revealed:

> Like you, I accepted the Copernicun [*sic*] position several years ago and discovered from
> thence the causes of many natural effects which are doubtless inexplicable by the current the-

ories. I have written up many of my reasons and refutations on the subject, but I have not dared until now to bring them into the open, being warned by the fortunes of Copernicus himself, our master, who procured immortal fame among a few but stepped down among the great crowd (for the foolish are numerous), only to be derided and dishonored. I would dare publish my thoughts if there were many like you; but, since there are not, I shall forebear....

In spite of his youthful exuberance, we can see in this last line of the letter to Kepler that Galileo was cautious about making public his belief in the Copernican system. Years later while in the midst of his work with the telescope, Galileo revealed his high esteem for Kepler* in a letter to his long-time correspondent: "You are the first and almost the only person who, even after but a cursory investigation, has, such is your openness of mind and lofty genius, given entire credit to my statement."

Galileo continued in the letter to Kepler, revealing the disdain he felt for the Aristotelian philosophers who disputed his discoveries and what they meant:

We will not trouble ourselves about the abuse of the multitude, for against Jupiter even giants, to say nothing of pigmies, fight in vain. Let Jupiter stand in the heavens, and let the sycophants bark at him as they will.... In Pisa, Florence, Bologna, Venice, and Padua many have seen the planets; but all are silent on the subject and undecided, for the greater number recognize neither Jupiter nor Mars and scarcely the moon as planet. At Venice one man spoke against me, boasting that he knew for certain that my satellites of Jupiter, which he had several times observed, were not planets because they were always to be seen with Jupiter, and either all of or some of them, now followed and now preceded him.

Greek traditions arising from the philosophy of Plato dictated a deep suspicion of sensory data, especially those collected through sight. Couple this with the fact that Galileo's discoveries employed an instrument whose basic physical properties were not understood, and the philosophical community had extra cause to doubt the reliability of his observations. Galileo, however, was particularly disdainful of the faculty of the leading universities, even his own, who preferred rhetoric and appeal to ancient authorities over observation and evidence.

I think, my Kepler, we will laugh at the extraordinary stupidity of the multitude. What do you say to the leading philosophers of the faculty here, to whom I have offered a thousand times of my own accord to show my studies, but who with the lazy obstinacy of a serpent who has eaten his fill have never consented to look at planets, nor moon, nor telescope? Verily, just as serpents close their ears, so do these men close their eyes to the light of truth. These are great matters; yet they do not occasion any surprise. People of this sort think that philosophy is a kind of book like the Aeneid or the Odyssey, and that the truth is to be sought, not in the universe, not in nature, but (I use their own words) *by comparing texts!* How you would laugh if you heard what things the first philosopher of the faculty at Pisa brought against me in the presence of the Grand Duke, for he tried, now with logical arguments, now with magical adjurations, to tear down and argue the new planets out of heaven.

Several years later, an embattled Galileo echoed these same sentiments in a letter to Christina, the grand duchess of Tuscany, and a member of the powerful Medici family.

Some years ago, as Your Serene Highness well knows, I discovered in the heavens many things that had not been seen before our own age. The novelty of these things, as well as

*In spite of his esteem for Kepler, Galileo never accepted the German astronomer's claim that the planets orbited the Sun on elliptical paths. Galileo continued to cling to the ancient belief in circular motion in the heavens.

some consequences which followed from them in contradiction to the physical notions commonly held among academic philosophers, stirred up against me no small number of professors — as if I had placed these things in the sky with my own hands in order to upset nature and overturn the sciences. They seemed to forget that the increase of known truths stimulates the investigation, establishment, and growth of the arts; not their diminution or destruction. Showing a greater fondness for their own opinions than for truth, they sought to deny and disprove the new things which, if they had cared to look for themselves, their own senses would have demonstrated to them.

In the years immediately after his telescopic discoveries — and the ensuing defense of Copernican astronomy — Galileo's troubles stemmed primarily from attacks from the scholarly establishment, not the church. This, however, was soon to change.

Science Versus the Church: The Galileo Affair

In the year sixteen hundred and nine
Science's light began to shine.
At Padua City, in a modest house,
Galileo Galilei set out to prove
The sun is still, the earth is on the move.

These lines, penned by Bertolt Brecht in his play *Galileo*, reduce to a simple verse the observations that set in motion a chain of events that remain today one of the best known episodes in the history of science — Galileo's condemnation by the Inquisition and the Catholic Church. How did Galileo, chief philosopher and mathematician to one of the most powerful men in the world and friend to cardinals and even popes, end up imprisoned and alone, his reputation (at least among the devout) ruined?

Although there had certainly been attacks from various opponents in the name of orthodox religion, Galileo's first direct encounter with the church over his beliefs occurred in 1615 when he traveled to Rome to defend his Copernican views. The following year, Galileo was instructed by Cardinal Bellarmine, a highly placed official at the Vatican, not to hold or defend the Copernican theory. In the same year, Copernicus' *On the Revolutions* was put on the Index of Prohibited Books, placing anyone who defended the book on dangerous ground. However, in an audience with Pope Paul V, Galileo was assured that he had not been condemned by the Inquisition.

In a move that should have boded well for Galileo, his friend Cardinal Barberini became Pope Urban VIII in 1623. The following year, in yet another audience with the pope, Galileo was assured that he could write about Copernicanism as a mathematical hypothesis as long as he did not teach it as physical truth. Encouraged by what he perceived as support from the church, Galileo began writing a book comparing the Copernican system to the Ptolemaic. Galileo chose to compose this work in the form of a dialogue between three acquaintances: Salviati, who defended the Copernican view; Simplicio, a staunch Aristotelian who defended the Ptolemaic system; and Sagredo, who is ostensibly neutral and often acted as a mediator between the other two men.

By presenting both world views, Galileo technically fulfilled the requirements that he present Copernicanism as only one theory among others. However, it becomes quite clear as the dialogue progresses that Salviati is winning each argument, and in the process making Simplicio seem foolish in defending the older system. In fact, Galileo himself admitted in the preface what he was about to do:

> To this end I have taken the Copernican side in the discourse, proceeding as with a pure mathematical hypothesis and striving by every artifice to represent it as superior to supposing the earth motionless–not, indeed absolutely, but as against the arguments of some professed Peripatetics.

These Peripatetics* always drew the scorn of Galileo:

> These men indeed deserve not even that name, for they do not walk about; they are content to adore the shadows, philosophizing not with due circumspection but merely from having memorized a few ill-understood principles.

Once he finally took a firm stand on Copernicanism, the attacks on Galileo intensified from philosophers and clergy alike. Bertolt Brecht imagined this conversation between monks and scholars discussing Galileo's adoption of Copernicanism and the radical idea that the earth moves:

> A Monk: It's rolling fast, I'm dizzy. May I hold on to you, Professor?
>
> The Scholar: Old Mother Earth's been at the bottle again. Whoa!
>
> Monk: Hey! Hey! We're slipping off! Help!
>
> Second Scholar: Look! There's Venus! Hold me, lads. Whee!
>
> Second Monk: Don't, don't hurl us off onto the moon. There are nasty sharp mountain peaks on the moon, brethren!
>
> Variously: Hold tight! Hold tight! Don't look down! Hold tight! It will make you giddy!
>
> Fat Prelate: And we cannot have giddy people in holy Rome.

Although seemingly light hearted banter, the underlying venom with which these lines are delivered certainly matches how much both scholars and monks despised Galileo and his theories. Brecht also imagined a more serious attack on Galileo's ideas from an old cardinal:

> So you have degraded the earth despite the fact that you live by her and receive everything from her. I won't have it! I won't have it! I won't be a nobody on an inconsequential star briefly twirling hither and tither. I tread the earth, and the earth is firm beneath my feet, and there is no motion to the earth, and the earth is the center of all things, and I am the center of the earth, and the eye of the Creator is upon me. About me revolve, affixed to their crystal shells, the lesser lights of the stars and the great light of the sun, created to give light upon me that God might see me — Man, God's greatest effort, the center of creation.

This much more serious accusation, that Galileo had removed humans from the center of God's creation, made the heliocentric theory and the idea that the earth moved seem more than ludicrous — it was blasphemous.

Galileo finished *Dialogue Concerning the Two Chief World Systems* in 1630 and began the lengthy process of gaining approval to publish through the church censors. Upon receiving the conditional approval of the secretary of the Vatican, *Dialogue* was printed and distributed. However, in 1632 Pope Urban stopped the distribution of the book and

---

*Peripatetics, from the Greek meaning "walking around," was the name given to the school of philosophers who followed Aristotle.

summoned Galileo to Rome to appear before him and the Inquisition. After some delay due to the health of Galileo (a delay that the pope suspected was contrived by Galileo), he finally appeared before the Inquisition in 1633. Galileo was convicted, forced to abjure, and sentenced to house arrest. Galileo spent the rest of his life under house arrest, with his coming and going as well as his contacts with the world severely restricted by the church officials. Galileo did take advantage of this forced seclusion by going back to his work on physics, and was able to complete his masterpiece in this field.

In spite of the devastating outcome of Galileo's trial before the Inquisition, it is clear that Galileo, a devout Catholic all of his life, never intended anything heretical in his writings or teachings. Many years before his trial, Galileo had already made clear his views on science and religion in letters sent to friends and patrons. One such letter, written in 1613, was addressed to Benedetto Castelli. Castelli, initially a student of Galileo, had become a great defender of both Galileo and the Copernican system. In the letter, Galileo discusses his belief that both nature and scripture have the same author, and when they seem to conflict, it is the interpretation of scripture that should be reexamined:

> As therefore, the Holy Scriptures in many places not only admit but actually require a different explanation for what seems to be the literal one, it seems to me that they ought to be reserved for the last place in mathematical discussions. For they, like nature, owe their origin to the Divine Word; the former is inspired by the Holy Spirit, the latter as the fulfillment of the Divine commands; it was necessary, however in Holy Scripture, in order to accommodate itself to the understanding of the majority, to say many things which apparently differ from the precise meaning.

Galileo contrasted the interpretation of scripture with a scientist's reading of nature:

> Nature, on the contrary, is inexorable and unchangeable, and cares not whether her hidden causes and modes of working are intelligible to the human understanding or not, and never deviates on that account from her prescribed laws.

He goes on to argue that passages of scripture "contain thousands of words admitting of various interpretations," and therefore no effects of nature "should be rendered doubtful by passages of Scripture."

So, Galileo continued, what is one to do if these two holy books, the book of nature and the book of God, appear to contradict?

> Since two truths can obviously never contradict each other, it is the part of wise interpreters of Holy Scripture to take the pains to find out the real meaning of its statements, in accordance with the conclusions regarding nature which are quite certain, either from the clear evidence of sense or from necessary demonstration. As therefore the Bible, although dictated by the Holy Spirit, admits, from the reasons given above, in many passages of an interpretation other than the literal one; and as, moreover, we cannot maintain with certainty that *all* interpreters are inspired by God, I think it would be the part of wisdom not to allow anyone to apply passages of Scripture in such a way as to force them to support, as true, conclusions concerning nature the contrary of which may afterwards be revealed by the evidence of our senses or by necessary demonstration.

Galileo then asks the dangerous question: "Who will set bounds to man's understanding? Who can assure us that everything that can be known in the world is already known?"

Of course, the church leaders would be the safe answer. For centuries, it was these men who had decided on all questions of faith as well as science, morality, politics, etc. for the faithful. Galileo, however, makes a different proposal:

> It would therefore perhaps be best not to add, without necessity, to the articles of faith which refer to salvation and the defense of holy religion, and which are so strong that they are in no danger of having at any time cogent reasons brought against them, especially when the desire to add to them proceeds from persons who, although quite enlightened when they speak under Divine guidance, are obviously destitute of those faculties which are needed, I will not say for the refutation, but even for the understanding of the demonstrations by which the higher sciences enforce their conclusions.

Galileo, in a not so subtle way, was making his case that the church fathers were not necessarily qualified to refute scientific discoveries, as they may not even have the facilities to understand the higher sciences. A dangerous assertion indeed! Galileo began his conclusion to the letter to Castelli by reasserting the proper places for scripture and science:

> I am inclined to think that the authority of Holy Scripture is intended to convince men of those truths which are necessary for their salvation, and which being far above man's understanding cannot be made credible by any learning, or any other means than revelation by the Holy Spirit. But that the same God has endowed us with senses, reason, and understanding, does not permit us to use them, and desires to acquaint us in any other way with such knowledge as we are in a position to acquire for ourselves by means of those faculties, *that* it seems to me I am not bound to believe.

Galileo refined and sharpened his ideas concerning apparent conflict between scientific observation and Holy Scripture in a letter to Christina, grand duchess of Tuscany. Christina was the mother of Cosimo (II), Galileo's most prominent patron. She was also very fond of Galileo and on more than one occasion had drawn him into serious discussions concerning the implication of his discoveries on theology. In this letter, composed in 1615, Galileo presented a classic and eternal apologetic for scientific inquiry and its quest to gain independence from religion. Referring to the philosophers who attacked his work, Galileo wrote:

> To this end they hurled various charges and published numerous writings filled with vain arguments, and they made the grave mistake of sprinkling these with passages taken from places in the Bible which they had failed to understand properly, and which were ill suited to their purposes. These men would perhaps not have fallen into such error had they but paid attention to a most useful doctrine of St. Augustine's, relative to our making positive statements about things which are obscure and hard to understand by means of reason alone. Speaking of a certain physical conclusion about the heavenly bodies, he wrote: "Now keeping always our respect for moderation in grave piety, we ought not to believe anything unadvisedly on a dubious point, lest in favor to our error we conceive a prejudice against something that truth hereafter may reveal to be not contrary in any way to the sacred books of either the Old or the New Testament."

Galileo continued by once again making a shocking proposal — answers provided by science should prevail over those provided by scripture when addressing physical problems in the universe.

> This being granted, I think that in discussions of physical problems we ought to begin not from the authority of scriptural passages, but from sense-experiences and necessary demon-

strations; for the holy Bible and the phenomena of nature proceed alike from the divine Word, the former as the dictate of the Holy Ghost and the latter as the observant executrix of God's commands. It is necessary for the Bible, in order to be accommodated to the understanding of every man, to speak many things which appear to differ from the absolute truth so far as the bare meaning of the words is concerned. But Nature, on the other hand, is inexorable and immutable; she never transgresses the laws imposed upon her, or cares a whit whether her abstruse reasons and methods of operation are understandable to men. For that reason it appears that nothing physical which sense-experience sets before our eyes, or which necessary demonstrations prove to us, ought to be called in question (much less condemned) upon the testimony of biblical passages which may have some different meaning beneath their words. For the Bible is not chained in every expression to conditions as strict as those which govern all physical effects; nor is God any less excellently revealed in Nature's actions than in the sacred statements of the Bible. Perhaps this is what Tertullian meant by these words: "We conclude that God is known first through Nature, and then again, more particularly, by doctrine; by Nature in His works, and by doctrine in His revealed word."

And finally, Galileo concluded, the current process of requiring science to bend to the interpretation of scripture is the exact reverse of how things should be: "This granted, and it being true that two truths cannot contradict one another, it is the function of wise expositors to seek out the true senses of scriptural texts."

Galileo: Physicist

Although Galileo is best remembered today (at least among non-scientists) for his astronomy and for his trial by the Inquisition, his contributions to natural philosophy, or physics, are equally important. For centuries, science had been dominated by the logico–deductive system of Aristotle and the ancient Greeks. Galileo introduced a new way of approaching the study of nature with a combination of experimental evidence and mathematical proof. In fact, Galileo was pleased to find that his experiments actually agreed with the mathematical derivations. In his *Discourse on Two New Sciences*, published in 1638, Galileo wrote: "In this belief we are confirmed mainly by the consideration that experimental results are seen to agree with and exactly correspond with those properties which have been, one after another, demonstrated by us."

While under house arrest (which, as it turns out, was to last the remainder of Galileo's life), the author of *The Starry Messenger* and *Dialogue Concerning the Two Chief World Systems* turned his attentions back to the work of his earlier years. In 1590, Galileo had prepared a manuscript — never published — he called *On Motion* in which he began to explore the science and mathematics of bodies in motion. Returning to these interests, Galileo produced *Discourse on Two New Sciences* in which he laid out his theories and experimental results concerning motion and material strength.

Galileo had long held that mathematics was the key to understanding nature. In 1623, in a work called the *Assayer*, he maintained:

Philosophy is written in this grand book — I mean the universe — which stands continually open to our gaze, but it cannot be understood unless one first learns to comprehend the language and interpret the characters in which it is written. It is written in the language of mathematics, and its characters are triangles, circles, and other geometrical figures, without which it is humanly impossible to understand a single word of it; without these, one is wandering around in a dark labyrinth.

So a mathematical exploration of motion and materials from the pen of Galileo completed a lifetime of work combining experimentation with mathematics.

We should recall, in Aristotelian physics an object's motion was attributed to the object seeking its natural place. Heavy elements (earth and water) fell towards the earth, while light elements (air and fire) rose above the earth. The speed of a falling body, according to Aristotle, was proportional to its weight. In other words, a ten pound ball should fall to the ground ten times faster than a one pound ball. This physics went almost unchallenged for centuries. But, even before Galileo, a few skeptics did arise to challenge Aristotle's theories on motion. Galileo, following in the footsteps of these few medieval mathematicians, took up the challenge of creating a new understanding of the cause of motion. In *Discourse*, Galileo employed the same rhetorical device he had previously used in *Dialogue Concerning the Two Chief World Systems*, namely a dialogue between three friends (in fact, the same three friends). In one section, the three discuss Aristotle's views on the motion of falling bodies:

> SIMPLICIO: He [*Aristotle*] supposes bodies of different weight to move in one and the same medium with different speeds which stand to one another in the same ratio as the weights; so that, for example, a body which is ten times as heavy as another will move ten times as rapidly as the other.

> SALVIATI: I greatly doubt that Aristotle ever tested by experiment whether it be true that two stones, one weighing ten times as much as the other, if allowed to fall, at the same instant, from a height of, say, 100 cubits, would so differ in speed that when the heavier had reached the ground, the other would not have fallen more than 10 cubits.

> SAGREDO: But I, Simplicio, who have made the test can assure you that a cannon ball weighing one or two hundred pounds, or even more, will not reach the ground by as much as a span ahead of a musket ball weighing only half a pound, provided both are dropped from a height of 200 cubits.

Galileo continued the discussion by proposing a thought experiment which led to a logical inconsistency in Aristotle's theory:

> SALVIATI: If then we take two bodies whose natural speeds are different, it is clear that on uniting the two, the more rapid one will be partly retarded by the slower, and the slower will be somewhat hastened by the swifter. Do you not agree with me in this opinion?

> SIMPLICIO: You are unquestionably right.

> SALVIATI: But if this is true, and if a large stone moves with a speed of, say, eight while a smaller moves with a speed of four, then when they are united, the system will move with a speed less than eight; but the two stones when tied together make a stone larger than that which before moved with a speed of eight. Hence the heavier body moves with less speed than the lighter; an effect which is contrary to your supposition. Thus you see how, from your assumption that the heavier body moves more rapidly than the lighter one, I infer that the heavier body moves more slowly.

So Salviati led Simplicio into a logical contradiction arising from Aristotelian theories on free falling bodies. By tying two stones together, the lighter stone should slow down the heavier stone, making the total falling time less than that of the heavier stone alone. On the other hand, binding these two stones together make a weight that is greater than the heavier stone alone, so they should fall faster — a mind-boggling contradiction, at least for Simplicio, who replies "I am all at sea."

Did Galileo actually climb to the top of the Leaning Tower of Pisa, as the legend goes, and drop objects of different weights to see how fast they fell? Most historians agree that he did not perform this experiment, but rather proposed it as a thought experiment for which he was sure of the outcome. To accurately time an experiment such as dropping objects from great heights was beyond the capabilities of the technology Galileo had at his disposal. Instead, he slowed down the process by conducting experiments with pendulums and with rolling balls on inclined planes.

Galileo described mathematically the movement of a free-falling body — the distance traversed was proportional to the square of the time it fell. He arrived at this conclusion based on the assumption that a free-falling body moved with constant acceleration. Galileo also stated that a body moving along a smooth, level surface would continue to move in the same direction and at a constant speed until something acted to slow it down. This is essentially a statement of the law of inertia, which became a cornerstone of classical physics when it was reworked and restated by Isaac Newton as his first law of motion.

Galileo also derived a mathematical description of the path of a free-falling body, or projectile. He demonstrated that this path was a parabola, a claim Galileo made during the third day of the *Discourse*:

> It has been observed that missiles and projectiles describe a curved path of some sort; however no one has pointed out the fact that this path is a parabola. But this and other facts, not few in number or less worth knowing, I have succeeded in proving; and what I consider more important, there have been opened up to this vast and most excellent science, of which my work is merely the beginning, ways and means by which other minds more acute than mine will explore its remote corners.

Such a discovery, inconsequential as it may seem to the reader, laid the foundation for a new science of ballistics in which the path of projectiles (cannonballs, artillery, missiles, etc.) could be calculated and predicted with extremely impressive accuracy.

Galileo Galilei would be remembered as one of the most important scientists in history for either his discoveries with the telescope (and his subsequent support of Copernicus' heliocentric cosmos) *or* his experiments and groundbreaking conclusions in physics. That both came from the mind, and the pen, of one man makes Galileo, as some have called him, the father of modern science.

PRIMARY SOURCES

Galilei, Galileo. *The Assayer*, translated by Stillman Drake. *Discoveries and Opinions of Galileo*. New York: Doubleday Anchor, 1957.

_____. *Dialogue Concerning the Two Chief World Systems*, translated by Stillman Drake. Berkeley: University of California Press, 1953.

_____. *Dialogues Concerning Two New Sciences*, translated by Henry Crew and Alfonso de Salvio. New York: Macmillan, 1914.

_____. "Letter to Castelli," translated by Maurice Finocchiaro. *The Galileo Affair*. Berkeley: University of California Press, 1989.

_____. "Letter to the Grand Duchess Christina," translated by Stillman Drake. *Discoveries and Opinions of Galileo*. New York: Doubleday Anchor, 1957.

_____. *The Starry Messenger*, translated by Stillman Drake. *Discoveries and Opinions of Galileo*. New York: Doubleday Anchor, 1957.

Other Sources

Biagioli, Mario. *Galileo, Courtier: The Practice of Science in the Culture of Absolutism*. Chicago: University of Chicago Press, 1993.

Blackwell, Richard J., and Paolo Antonio Foscarini. *Galileo, Bellarmine, and the Bible: Including a Translation of Foscarini's Letter on the Motion of the Earth*. Notre Dame, IN: University of Notre Dame Press, 1991.

Brecht, Bertolt. *Galileo*. New York: Grove Press, 1966.

Clavelin, Maurice. *The Natural Philosophy of Galileo; Essay on the Origins and Formation of Classical Mechanics*. Cambridge: MIT Press, 1974.

Drake, Stillman. *Discoveries and Opinions of Galileo*. New York: Anchor, 1957.

_____. *Galileo*. New York: Hill and Wang, 1980.

_____. *Galileo at Work: His Scientific Biography*. Chicago: University of Chicago Press, 1978.

Finocchiaro, Maurice A. *Retrying Galileo, 1633–1992*. Berkeley: University of California Press, 2005.

Freedberg, David. *The Eye of the Lynx: Galileo, His Friends, and the Beginnings of Modern Natural History*. Chicago: University of Chicago Press, 2002.

Machamer, Peter K. *The Cambridge Companion to Galileo*. Cambridge and New York: Cambridge University Press, 1998.

MacLachlan, James H. *Galileo Galilei: First Physicist*. New York: Oxford University Press, 1997.

Moss, Jean Dietz. *Novelties in the Heavens: Rhetoric and Science in the Copernican Controversy*. Chicago: University of Chicago Press, 1993.

Newton, Roger G. *Galileo's Pendulum: From the Rhythm of Time to the Making of Matter*. Cambridge: Harvard University Press, 2004.

Redondi, Raymond. *Galileo: Heretic*. Translated by Raymond Rosenthal. Princeton: Princeton University Press, 1987.

Reston, James. *Galileo: A Life*. New York: HarperCollins, 1994.

Rowland, Wade. *Galileo's Mistake: A New Look at the Epic Confrontation Between Galileo and the Church*. New York: Arcade, 2003.

Shea, William R., and Mariano Artigas. *Galileo in Rome: The Rise and Fall of a Troublesome Genius*. New York: Oxford University Press, 2003.

Sobel, Dana. *Galileo's Daughter: A Historical Memoir of Science, Faith, and Love*. New York: Walker, 1999.

Wallace, William A. *Galileo and His Sources: The Heritage of the Collegio Romano in Galileo's Science*. Princeton: Princeton University Press, 1984.

CHAPTER 3

Copernicus Perfected?
Kepler and Planetary Orbits

But my exhausting task was not complete: I had a fourth step yet to make towards the physical hypotheses. By most laborious proofs and by computations on a very large number of observations, I discovered the course of a planet in the heavens is not a circle, but an oval path, perfectly elliptical. — Johannes Kepler, *New Astronomy* (1609)

Copernicus' insistence on the perfection of circular orbits not only motivated his search for a better cosmological theory but also inhibited him (and others) from a full understanding of the mechanics of a heliocentric system. It was not until the German mathematician and astronomer (and confirmed Copernican) Johannes Kepler realized that planetary orbits were not circles that the heliocentric cosmos of Copernicus was completed. Before we look into the life and work of Kepler, we must first begin with an astronomer of the generation before Kepler — an astronomer whose insistence on careful and accurate observations of the sky had a direct and long lasting influence on Kepler in particular, and astronomy in general.

Tycho Brahe

Tycho Brahe (1546–1601) was born into Danish nobility a mere three years after Copernicus' death and the appearance of *On the Revolutions*. Although never a full convert to the heliocentric system, Tycho's work cast doubts upon the traditional views of the universe, and the immense number of observations left by the Danish astronomer was the key used by Kepler to unlock the secrets of planetary orbits.

Tycho's father had promised his brother, who was childless, that he could have one of his own sons to raise as his own. After failing to live up to this promise, the uncle kidnapped the young boy much to his father's outrage. Eventually, the relationship was mended and Tycho became heir to his uncle's fortune. Tycho's interest in astronomy bloomed while studying at the University of Copenhagen, and he decided to dedicate his life to careful and systematic observations of the night sky. He continued his studies in astronomy at the University of Leipzig (Germany), against the wishes of both his uncle and his parents, who wanted him to study law.

As a young man of 20, Tycho became involved in a duel with swords and lost part of his nose. For the rest of his life, he wore an ornamental replacement, made of gold, silver, or copper. Like Galileo, Tycho never formally married. He did have a common law wife with whom he lived his entire adult life and with whom he had eight children. After inheriting his uncle's wealth, Tycho himself became one of the wealthiest men in Denmark. He was also exceedingly eccentric, holding lavish parties at his castle where he kept a tame moose, which, according to contemporary stories, became drunk on beer and died from a fall down the stairs.

Tycho used his wealth to build his own observatory and to furnish it with equipment to observe the night sky. In these days before the invention of the telescope, Tycho's equipment included devices such as quadrants, spheres, and various naked eye devices for observing and measuring celestial objects.

The year 1572 marked the beginning of Tycho's rise to fame as an astronomer. In November of that year, Tycho related:

> I was contemplating the stars in a clear sky, [when] I noticed that a new and unusual star, surpassing the other stars in brilliancy, was shining almost directly above my head....
>
> I was so astonished at this sight that I was not ashamed to doubt the trustworthiness of my own eyes. But when I observed that others, too, on having the place pointed out to them, could see that there was really a star there, I had no further doubts.

Tycho proceeded to summarize some centuries-old assumptions about the eternity of the heavens. At the same time, he seems to prepare his readers for his own astonishing conclusions:

> For all philosophers agree, and facts clearly prove it to be the case, that in the ethereal region of the celestial world no change, in the way either of generation or of corruption, takes place; but that the heavens and the celestial bodies in the heavens are without increase or diminution, and that they undergo no alteration, either in number, or in size, or in light or in any other respect; that they always remain the same, like unto themselves in all respects, no years wearing them away. Furthermore, the observations of all founders of the science, made some thousands of years ago, testify that all the stars have always retained the same number, position, order, motion, and size as they are found, by careful observation on the part of those who take delight in heavenly phenomena, to preserve even in our own day.

In spite of his summary of the ancient position on the immutability of the heavens, Tycho laid the foundation for some doubt by pointing out that Hipparchus had claimed to observe a new star in the heavens in the second century B.C.

Most astronomers and philosophers of the day would have left it at this; obviously the new star was either an apparition — some sort of atmospheric optical illusion — or an object between the earth and the moon where change did occur. Tycho, however, was unwilling to adhere to the status quo. Instead, he related: "In order, therefore, that I might find out in this way whether the star was in the region of the Element, or among the celestial orbits, and what its distance was from the earth itself, I tried to determine whether it had a parallax, and, if so, how great a one."

Tycho proceeded to provide more detailed observations and calculations concerning the actual location of this new star. He presented his geometrical proof that the star does not exhibit parallax and

> therefore, this new star is neither in the region of the Element, below the moon, nor among the orbits of the seven wandering stars, but it is in the eighth sphere, among the other fixed stars, which was what we had to prove. Hence it follows that it is not some peculiar kind of comet or some other kind of fiery meteor become visible.

Tycho's claim that the object was not a comet is based on two things. His own observations showed that the form of the new star was much different than that of a comet. In addition, Tycho attested to the fact that this star was in the outer heavens whereas all comets, agreed the philosophers, occured inside the moon's orbit:

> For none of these [comets] are generated in the heavens themselves, but they are below the moon, in the upper region of the air, as all philosophers testify; unless one would believe with Albategnius that comets are produced, not in the air, but in the heavens. For he believes that he has observed a comet above the moon, in the sphere of Venus. That this can be the case, is not yet clear to me. But, please God, sometime, if a comet shows itself in our age, I will investigate the truth of the matter.

Tycho's observation of the new star, and his careful placement of the object in the realm of the fixed stars, dealt a blow to the Aristotelian notion of the immutability of the heavens. Furthermore, his prayer was answered five years later when Tycho observed a comet and was able to place it in the realm of the planets. This confirmed what Tycho already strongly suspected in 1572 — changes do occur in the heavenly realm beyond the moon.

In fact, Tycho's calculations of the orbit of the comet of 1577 dealt another blow to the cosmology of Aristotle and Ptolemy. Since the ancient Greeks, philosophers had agreed that the planets were carried around on their orbits around the earth by crystalline spheres. Copernicus adopted the same assumption to explain what imparted motion to the planets. Tycho's observations showed the comet crossing the boundaries supposedly formed by these solid spheres. These observations metaphorically shattered the crystalline spheres.

The fame that Tycho's work brought him also served to keep him in Denmark. Although he had his own observatory from which he worked, Tycho contemplated leaving Denmark. King Frederick II had other plans. To entice Tycho to stay, Frederick granted him the island of Hven and generous funding to build an elaborate observatory and alchemical laboratory that Tycho called Uraniborg. Much more than an observatory, Uraniborg became a center of research under Tycho's direction with multitudes of assistants and students who came to train and learn under the now famous astronomer. Tycho designed and built the largest and most accurate instruments in the world. He also installed his own printing press from which the discoveries of his research institute could be published.

While at Uraniborg, Tycho also developed his own ideas concerning the makeup of the universe. The Tychonian system, as it came to be known, was geo-heliocentric. It had the sun and moon orbiting the central earth, while all of the other planets orbited the sun. This provided, in Tycho's estimation, for the best of both worlds, Copernican and Ptolemaic. Tycho could not accept the Copernican heliocentric model because he could not accept the idea of a moving earth. His rejection of the mobility of the earth was based on two arguments: one was the ancient belief that the earth was fundamentally different — heavy and sluggish and incapable of such movement. A heavy unmoving earth was central to Aristotelian physics which relied on the notion of natural motion with heavy objects seeking their natural place: a central earth. Although Copernicus provided a new cosmological theory that was mathematically sound and in many ways logically superior to the old geocentric system, he did not provide a viable replacement for Aristotelian physics. Tycho also resisted the Copernican system because he could not measure the required stellar parallax that would come with the acceptance of Copernicus.

One rather odd component of the Tychonian system was that the orbits of Mars and the sun intersected. Among many other difficulties, this meant that the Aristotelian concept that the planets were carried by crystalline spheres could not be true. When Tycho showed that the comet of 1577 crossed the paths of the planets, this confirmed, at least to him, that the crystalline spheres did not exist. The Tychonian system, and other similar systems involving geo-heliocentricity, survived for some time throughout Europe before finally succumbing to overwhelming evidence for the pure heliocentric cosmos.

Tycho spent many productive years working on his island observatory. However, by 1597 his circumstances had changed. His patron, King Frederick, was dead and his successor did not support Tycho's work in the way he was accustomed. As a renowned astronomer, Tycho had several different opportunities. He chose to move to Prague at the invitation of the Holy Roman Emperor Rudolph II where he became the imperial mathematician charged with casting horoscopes and advising Rudolph and his court based on astrological signs. Tycho was also given the funds to build a new observatory where he picked up where he had left off at Uraniborg. In 1600, after several letters of correspondence, Tycho hired a new assistant at his observatory in Prague. This new assistant was a talented young mathematician and astronomer from Germany named Johannes Kepler. Kepler's application of his superior mathematical abilities to the mountains of observations accumulated by Tycho over his long career led to some of the most important results in the history of science.

Johannes Kepler

Johannes Kepler (1571–1630) was a German-born Lutheran whose father was a soldier of fortune and died while Johannes was young. His mother was a healer whose trial for witchcraft after Johannes had already become an established astronomer and mathematician caused a great deal of concern, not to mention years of appeals and court appearances by Johannes in an ultimately successful attempt to vindicate his mother.

Unlike Tycho Brahe, Kepler's family had little money or reputation; it was only Kepler's precociousness that attracted the attention of teachers and paved the way for financial assistance so that the young Johannes might apply his abilities in mathematics. Although a bout with smallpox affected his eyesight and ruined his abilities as an observational astronomer, Kepler's interest in mathematical astronomy never waned. Kepler developed a deep Platonic philosophy that God had created a world best understood through mathematics. His lifelong goal was to discover the underlying mathematical secrets of this world and thus to move closer to God.

Kepler studied at the University of Tübingen. Although the official curriculum in astronomy was the traditional Ptolemaic geocentric system, he was taught by Michael Maestlin, a leading astronomer and supporter of Copernicanism. Kepler was recognized as a brilliant mathematics student, and was soon appointed to teach at the University of Graz in Austria. This proved to be the first of many positions held by Kepler. Germany (as well as most of Europe) at the time was split by the flaming passions ignited by the Reformation. Kepler was often forced to move due to shifting religious (and political) allegiances; but even within Protestant states, Kepler's refusal to accept some Lutheran orthodoxies caused great consternation among officials and led to termination of his services.

In 1600, Kepler met Tycho Brahe at his observatory near Prague. Little did he know at the time that his life — and the history of science — would change forever. In spite of a sometimes contentious relationship, Kepler became Brahe's assistant. After only a brief period as his assistant, Brahe died unexpectedly and Kepler was named as his replacement as imperial mathematician to the holy Roman emperor. Much like Brahe, Kepler's chief duty was as an astrologer; unlike Brahe, however, Kepler had always maintained a fascination with astrology. Many publications on astrology throughout his life helped support Kepler in lean times. Although the most productive years of his professional life occurred in the service of Rudolph II in Prague, once again shifting political alliances (along with the death of his wife) led Kepler to move to Linz, and later Ulm, where he died in 1630.

Kepler's publications form perhaps the most intriguing body of work of any scientist in history. The important discoveries made by Kepler — in particular his three laws of planetary motion — for which we remember him today are only a part of the story. Kepler made important contributions to the science of optics, to the emerging branch of mathematics we now call calculus, and to the philosophy of science.

In the *Optics*, Kepler presented his theories on the nature of light and offered a new theory of vision in which he correctly surmised that images are inverted on the retina. Kepler's work in the science of optics also led him to develop an improved telescope, known as the Keplerian telescope.

His mind always at work, Kepler noticed at his own wedding celebration that the method used for measuring the volume of wine in barrels and casks of various sizes was inaccurate and misunderstood. Kepler developed a method, using techniques from the legendary Greek mathematician Archimedes, to more accurately estimate the amount of wine in any given barrel. His method employed the idea of infinitesimals and is considered a forerunner of integral calculus.

During his lifetime, Kepler's compilation of a complete and systematic set of astronomical tables (based on Brahe's years of observations) was considered his crowning achievement. He spent years preparing ephemerides, or predictions of planetary and stellar positions, and eventually named the resulting tables the Rudolphine Tables after his benefactor, Rudolph II. Kepler used logarithms in his development of the tables. Logarithms, recently invented by a Scot by the name of John Napier, were not widely accepted or used because they were not understood. Kepler carefully showed how logarithms worked and his tables were the most accurate and widely used for many years.

In addition to these many contributions to the development of modern science, Kepler published several books that contained theories that, although not considered important to science today, were every bit as important to his reputation as those theories that have withstood the test of time. He also wrote what many consider to be the first work of science fiction. In *The Dream*, Kepler traveled to the moon to investigate astronomy and cosmology from a vantage point other than the earth.

Kepler Wrestles with Mars: The Laws of Planetary Motion

Today, if the average person knows anything about Johannes Kepler, it is almost certain to be his first law of planetary motion: the orbits of the planets are elliptical with the sun at one focus of the ellipse. The story of Kepler's discovery of this law is one of the most fascinating stories in the history of science.

When Tycho Brahe died suddenly in 1601, Kepler not only assumed his position as imperial mathematician, he also gained some control (in spite of ongoing disputes with Brahe's heirs) of the voluminous data accumulated over a lifetime of observations. In order to better understand the orbits of the planets, and to provide evidence for the Copernican theory of the universe, Kepler began a painstaking mathematical analysis of the orbit of Mars. After years of tedious calculation — and numerous false theories and dead ends — Kepler finally concluded that the orbits were indeed elliptical, a conclusion that, he later wrote, came suddenly "as one aroused from sleep who gazed with astonishment on a new light!"

There are several reasons why it took Kepler so long to conclude that the orbits were elliptical. First, lest we forget, thousands of years of tradition and authority made it difficult, if not impossible, for most to even conceive of an orbit that was not circular. In his *New Astronomy*, published in 1609, Kepler wrote of the blinders worn by astronomers, including himself: "My first error was to suppose that the path of the planet is a perfect circle, a supposition that was all the more noxious a thief of time the more it was endowed with the authority of all philosophers, and the more convenient it was for metaphysics in particular."

It required a great leap and an original mind to even consider that these philosophers, from Aristotle to Copernicus, were in error. In addition to the metaphysical shackles worn by Kepler and all astronomers, it should be understood that the difference between the elliptical path followed by Mars (and most other planets) and a true circle is very small, especially when compared to the level of error found in the astronomical observations of the time. Drawn to scale, the ellipse generated by the orbit of Mars looks very much like a circle to the casual observer. If Kepler had chosen Mercury to study, for instance, whose elliptical path is more eccentric (meaning more stretched out), it might have been more immediately apparent that the circle was wrong.

This first law of planetary orbits first appeared in *New Astronomy*, along with his second law: the orbits of the planets sweep out equal areas in equal times. This book was a very mathematical work, as were all of Kepler's publications. Reflecting on the fact that the traditional Greek form of mathematics — state a proposition, explain its meaning, present a proof, and draw a conclusion — continued to dominate mathematical thinking, Kepler commented on the difficulty of composing such a work: "It is extremely hard these days to write mathematical books, especially astronomical ones. For unless one maintains the truly rigorous sequence of proposition, construction, demonstration, and conclusion, the book will not be mathematical; but maintaining the sequence makes the reading most tiresome."

Echoing Copernicus' famous line, "Mathematics is for mathematicians," Kepler concluded, "Morever, there are very few suitably prepared readers these days: the rest generally reject such works."

Again, reflecting Copernicus' disdain for those "philosophers" incapable, or unwilling, to accept the evidence presented, Kepler wrote:

> But whoever is too stupid to understand astronomical science, or too weak to believe Copernicus without affecting his faith, I would advise him that, having dismissed astronomical studies and having damned whatever philosophical opinions he pleases, he mind his own business and betake himself home to scratch in his own dirt patch, abandoning this wandering about the world.

In all of his astronomical treatises, Kepler defended the heliocentric cosmos of Copernicus. In the *New Astronomy* Kepler made several arguments, physical and metaphysical, for the sun's centrality. He wrote of the sun's "dignity" and its illumination. He also made an argument against the system of his old boss and mentor, Tycho Brahe, who had the sun moving around the earth while the other planets moved around the sun:

> Now let us consider the bodies of the sun and the earth, and decide which is better suited to being the source of motion for the other body. Does the sun, which moves the rest of the planets, move the earth, or does the earth move the sun, which moves the rest, and which is so many times greater? Unless we are forced to admit the absurd conclusion that the sun is moved by the earth, we must allow the sun to be fixed and the earth to move.

Furthermore, claimed Kepler, if there were no crystalline spheres, as Brahe had shown with his observation of the comet of 1577, what kept the planets in their place? The answer was the sun, with a force analogous to a magnet. To search for physical reasons for celestial motions was a new idea and not widely accepted — in fact it was strongly resisted by many of Kepler's peers.

It's difficult to comprehend the mathematical sophistication of Kepler's works without careful study of the details he included. Kepler related to his readers the difficulties of his work when he finally revealed his conclusion about elliptical orbits. Even after working through three exhausting and exhaustive mathematical steps to find the true nature of the movements of the planets, Kepler was not finished with his work:

> But my exhausting task was not complete: I had a fourth step yet to make towards the physical hypotheses. By most laborious proofs and by computations on a very large number of observations, I discovered the course of a planet in the heavens is not a circle, but an oval path, perfectly elliptical.

Laborious indeed! After many years of mind-numbing calculations, Kepler had finally arrived at what we now call the first law of planetary motion.

The culmination of Kepler's work came some years later in a multi–volume treatise known as the *Epitome of Copernican Astronomy*. This book was Kepler's crowning achievement in which he published all three laws of planetary motion and extended elliptical orbits from Mars to all planets and moons. He also continued to make his physical arguments for the motions of the heavens. Although written in a question and answer format to make it as easy as possible to follow, the book was still a work of mathematical astronomy.

In the *Epitome of Copernican Astronomy*, Kepler reiterated many of his theories, both mathematical and theoretical, which — at least to him — supported the Copernican thesis. For instance, when queried as to what holds the planets in their orbits around the sun, Kepler replied it is the power of the attraction and repulsion of magnetism:

> For repulsion and attraction take place according to the lines of virtue going out from the centre of the sun; and since those lines revolve along with the sun, it is necessary for the planet which is repelled and attracted to follow those lines in proportion to their strength in relation to the resistance of the planetary body. So the contrary movements of repulsion and attraction somehow compose this laying hold [*of the planet by the sun*].

As to the question, "*But isn't it unbelievable that the celestial bodies should be certain huge magnets?*" Kepler referred the reader to William Gilbert's work in which he teaches that the earth acts as a giant magnet (see Chapter 4). If, Kepler asked, the earth might behave in such a way, why not the other planets?

The title of this work being the *Epitome of Copernican Astronomy*, Kepler answered the very reasonable question, "*By what right do you make this* [elliptical orbits] *also a part of Copernican astronomy, since that author abided by the opinion of the ancients concerning perfect circles?*"

Kepler responded:

> I admit that this formulation of the hypothesis is not Copernican. But because the part concerning the eccentric circle is subordinate to the general hypothesis which employs the annual movement of the earth and the stillness of the sun: therefore the name comes from the more important part of the hypothesis.

From this statement we can see that the centrality of the sun, and the movement of the earth, was central to the Copernicanism adopted by Kepler. However, we must remember that to say that Copernicus was committed to circular motion is to understate his convictions. The idea of perfect circular motion was central to Copernicus' motivation for investigating a new theory of the universe. Circular motion was Copernicanism (at least to Copernicus), which is what prompts historian Richard Westfall to write, "If it is true that Kepler perfected Copernican astronomy, it is equally true that he destroyed it."

Secrets and Harmonies

Although we find most of Kepler's long-lasting influential ideas in the *New Astronomy* and the *Epitome of Copernican Astronomy*, two other works published by Kepler had a great deal of influence on his contemporaries and were considered by the author himself to be central to his life's work. Although many of the ideas in these books (*The Secret of the Universe* and *Harmonies of the World*) may seem foreign and even a little outlandish to the modern reader, their importance to seventeenth century astronomy — as well as what they reveal about the mind of Kepler and his contemporaries — make them worthy of study for the modern student of history.

Kepler published his first major work, *The Secret of the Universe*, in 1596 and it was immediately hailed by his contemporaries in mathematical astronomy. In this treatise, Kepler announced his discovery that the six known planets fit into orbits encased by five regular (Platonic) solids. The Platonic solids (the cube, the tetrahedron, the octahedron, the dodecahedron, and the icosahedron) are solids whose faces are congruent regular polygons. For instance, all of the faces of the cube are congruent squares; the faces of the tetrahedron are all congruent triangles; and so on. These solids held a special place in the mathematical mysticism of the Pythagoreans, and this importance had been preserved through the ages. Kepler's discovery that they were tied to the orbits of the planets seemed to confirm their special place in nature. The fact that this discovery only worked if the six planets orbited the sun was a confirmation to Kepler that the Copernican system was correct. Kepler believed that he had found the very blueprint, in the language of geometry, with which God had created the world. Years after his discovery, writing in the *Epitome of Copernican Astronomy*, when asked for the reason that there were exactly six planets, Kepler responded that it is due to geometric considerations.

> But among the rectilinear magnitudes the first, the most perfect, the most beautiful, and most simple are those which are called the five regular solids. More than 2,000 years ago Pythagoreans said that these five were the figures of the world, as they believed that the four elements and the heavens — the fifth essence — were conformed to the archetype for these five figures.

In other words, there are six planets because their orbits fit perfectly inside of the five Platonic solids.

Kepler opens the *Secret of the Universe* with a simple explanation of what he wished to accomplish:

GREETINGS, FRIENDLY READER

> The nature of the universe, God's motive and plan for creating it, God's source for the numbers, the law for such a great mass, the reason why there are six orbits, the spaces which fall between all the spheres, the cause of the great gap separating Jupiter and Mars, though they are not in the spheres — here Pythagoras reveals all this to you by five figures. Clearly he has revealed by this example that we can be born again after two thousand years of error, a better explorer of the universe. But hold back no longer from the fruits found within these rinds.

He then wrote:

> It is my intention, reader, to show in this little book that the most great and good Creator, in the creation of this moving universe, and the arrangement of the heavens, looked ... to those five regular solids, which have been so celebrated from the time of Pythagoras and Plato down to our own, and that he fitted to the nature of those solids, the number of heavens, their proportions, and the law of their motions.

Kepler's motivation was simple. All he wished to do was show how God created the universe and what tools he used in this creation!

Kepler eventually revealed to the reader the surprising simplicity of his discovery and the beautiful geometry of the universe:

> The earth is the circle which is the measure of all. Construct a dodecahedron round it. The circle surrounding that will be Mars. Round Mars construct a tetrahedron. The circle surrounding that will be Jupiter. Round Jupiter construct a cube. The circle surrounding that will be Saturn. Now construct an icosahedron inside the earth. The circle inscribed within that will be Venus. Inside Venus inscribe an octahedron. The circle inscribed within that will be Mercury.

Based partially on mathematical precepts and partially on metaphysical reasons, Kepler explained why the order of the solids which bound the orbits of each planet must be as they were:

> Now the cube should be close to the fixed stars, and establish the first proportion, that between Saturn and Jupiter, because the fixed stars are the most important part of the universe outside the earth.... Nobody will now greatly wonder why the pyramid [of which the tetrahedron is composed] follows the cube, since ... the former has almost dared to contend with the cube for the chief place.

Random as this may seem to the modern reader, Kepler's theory was actually the culmination of years of concentrated work and immense mathematical computation. Kepler's account of his eureka moment, when he first made his discovery, is an emotional description of the wonder of new knowledge:

> What delight I have found in this discovery I shall never be able to express in words. No longer did I regret the wasted time; I was no longer sick of the toil; I did not avoid the tedium of the calculation; I devoted my days and nights to computation, until such time as I could see whether the proposition which I had conceived in words would agree with the circles of Copernicus, or whether my joy would be scattered to the winds ... success came after a few days, and I found out how neatly one body fitted after another among the planets ... that you will have an answer for the peasant who asks what hooks the sky is hung on to prevent it from falling.

Even after his discovery of the three laws of planetary motion, and a lifetime of work that made him one of the preeminent astronomers of his generation (along with Galileo), Kepler continued to believe that his discovery of the geometric configuration of the universe was his crowning achievement.

When he wrote the *Secret of the Universe*, Kepler also introduced another of his theories in which he attempted to explain the structure of God's creation. Kepler was convinced that the universe not only represented geometrical harmony but also musical harmony. Commenting on the "kinship" between the musical harmonies and the Pythagorean solids, Kepler wrote: "Therefore the perfect harmonies must be fitted to the cube, pyramid, and octahedron, the imperfect to the dodecahedron and the icosahedron."

Kepler believed that the motion of the planets could be analyzed in terms of the musical notes they represented. He expanded on this idea in a later work he entitled the *Harmonies of the World*, published in 1619. Kepler's exploration of "the music of the spheres" was not new or unusual to him. The relationship between notes and harmonies had been studied as one of the mathematical arts since the time of the Pythagoreans, and Kepler, among others, understood it to be just one part of God's harmonious world. As part of this harmonious world, Kepler stated what became known as his third law of planetary motion. This law relates the period of a planet (the time it takes to complete one orbit around the sun) to the radius of its orbit. In particular, Kepler discovered that the square of the period is directly proportional to the cube of its radius.

Johannes Kepler was hailed during his lifetime as a brilliant mathematician and astronomer. His publications were well-received, and he served some of the most powerful men in Europe as court mathematician and astrologer. Kepler supported Copernicanism in a time when support of the heliocentric theory continued to be controversial, even dangerous. While Kepler was protected by powerful patrons sympathetic to his ideas, Galileo was being tried and convicted by the Inquisition for harboring the same ideas. Kepler's life, however, was not without controversy. He was forced to move several times due to changing political and religious realities, and his own unorthodox beliefs caused him trouble wherever he went.

Kepler — as well as most other Copernicans of the seventeenth century — did face opposition to his theories from the established academic community. In the *Epitome of Copernican Astronomy*, Kepler addressed this ingrained conservatism of the universities:

> In regards the academies, they are established in order to regulate the studies of the pupils and are concerned not to have the program of teaching change very often; in such places, because it is a question of the progress of the students, it frequently happens that the things which have to be chosen are not those which are most true but those which are most easy. And by that division in things which makes different people form different judgements, it so happens that certain people are in error contrary to their own opinion. It seems to me the truth concerning the mutable nature of the heavens can be taught conveniently; but someone else judges that students and teachers equally are thrown into confusion by this doctrine.

Kepler's mathematical explanations of the workings of the universe, along with his groundbreaking attempts to tie physical explanations to these mathematical theories, paved the way for future scientists to break the stranglehold of tradition on physical and mathematical astronomy.

Primary Sources

Brahe, Tycho. *The New Star*. John H. Walden, translator. *Philosophy of Science: An Historical Anthology*, Timothy J. McGrew, Marc Alspector-Kelley, and Fritz Allhoff, editors. Malden, MA: Blackwell, 2009.

Kepler, Johannes. *Epitome of Copernican Astronomy*. Charles Glenn Wallis, translator. *The Great Books*, Encyclopedia Britannica, 1952.

_____. *New Astronomy*. William H. Donahue, translator. Cambridge: Cambridge University Press, 1992.

_____. *The Secret of the Universe*. A.M. Duncan, translator. New York: Abaris Books, 1981.

Other Sources

Caspar, Max. *Kepler*. London and New York: Abelard-Schuman, 1959.

Christianson, J.R. *On Tycho's Island: Tycho Brahe and His Assistants, 1570–1601*. Cambridge and New York: Cambridge University Press, 2000.

_____. *On Tycho's Island: Tycho Brahe, Science, and Culture in the Sixteenth Century*. New York: Cambridge University Press, 2003.

Connor, James A. *Kepler's Witch: An Astronomer's Discovery of Cosmic Order Amid Religious War, Political Intrigue, and the Heresy Trial of His Mother*. San Francisco: HarperSanFrancisco, 2004.

Dreyer, J.L.E. *Tycho Brahe: A Picture of Scientific Work and Life in the Sixteenth Century*. New York: Dover, 1963.

Field, Judith Veronica. *Kepler's Geometrical Cosmology*. Chicago: University of Chicago Press, 1988.

Gingerich, Owen. *The Eye of Heaven: Ptolemy, Copernicus, Kepler.* New York: American Institute of Physics, 2003.

Hallyn, Fernand. *The Poetic Structure of the World: Copernicus and Kepler.* New York: Zone Books; Cambridge: Distributed by MIT Press, 1990.

Koestler, Arthur. *The Sleepwalkers: A History of Man's Changing Vision of the Universe.* New York: Macmillan, 1959.

_____. *The Watershed: A Biography of Johannes Kepler.* Garden City, NY: Anchor, 1960.

Koyre, Alexandre. *The Astronomical Revolution: Copernicus, Kepler, Borelli.* Paris: Hermann; Ithaca: Cornell University Press, 1973.

Kozhamthadam, Job. *The Discovery of Kepler's Laws: The Interaction of Science, Philosophy, and Religion.* Notre Dame, IN: University of Notre Dame Press, 1993.

Martens, Rhonda. *Kepler's Astronomy and the New Philosophy.* Princeton: Princeton University Press, 2000.

Mosley, Adam. *Bearing the Heavens: Tycho Brahe and the Astronomical Community of the Late Sixteenth Century.* Cambridge and New York: Cambridge University Press, 2007.

Rosen, Edward. *Three Imperial Mathematicians: Kepler Trapped Between Tycho Brahe and Ursus.* New York: Abaris, 1986.

Stephenson, Bruce. *Kepler's Physical Astronomy.* New York: Springer-Verlag, 1987.

_____, *The Music of the Heavens: Kepler's Harmonic Astronomy.* Princeton: Princeton University Press, 1994.

Thoren, Victor E., and J. R. Christianson. *The Lord of Uraniborg: A Biography of Tycho Brahe.* Cambridge and New York: Cambridge University Press, 1990.

Voelkel, James R. *The Composition of Kepler's Astronomia Nova.* Princeton: Princeton University Press, 2001.

CHAPTER 4

Gilbert, Harvey, and the Experimental Method

Clearer proofs, in the discovery of secrets, and in the investigation of the hidden causes of things, being afforded by trustworthy experiments and demonstrated arguments, than by the probably guesses and opinions of the ordinary professors of philosophy.—William Gilbert, *On the Magnet and Magnetic Bodies, and on the Great Magnet the Earth* (1600)

True philosophers, who are only eager for truth and knowledge, never regard themselves as already so thoroughly informed, but that they welcome further information from whomsoever and from wheresoever it may come; nor are they so narrow-minded as to imagine any of the arts or sciences transmitted to us by the ancients, in such a state of forwardness or completeness, that nothing is left for the ingenuity and industry of others.—William Harvey, *On the Motion of the Heart and Blood* (1628)

As we have seen, Renaissance natural philosophy was primarily dominated by Aristotle and his commentators. Aristotle's approach to science emphasized deduction over induction. In other words, observation of the natural world—and by extension experimentation—was not the central theme for the scientists of medieval and Renaissance Europe. Although there were Islamic philosophers of the Middle Ages who advocated for induction and experimentation, it was not until the sixteenth and seventeenth centuries that a few Europeans began to argue for empiricism; the experimental method in Western science was born.

Interestingly, the leader of the empirical method in early modern Europe is not remembered for any scientific discoveries or theories, but rather for his influential writings on the philosophy of science—in other words, his instructions to other philosophers on how science should be approached. Francis Bacon (1561–1626) was an English philosopher and politician from a well-connected family. He served the British throne in a variety of positions and was ultimately knighted in 1603. Unfortunately, Bacon was later caught up in a bribery scandal and was stripped of his titles and briefly imprisoned in the Tower of

London. He died of pneumonia contracted while experimenting with the preservation of meat by freezing.

Bacon introduced his philosophy of science in a work he called *Novum Organum*, or *The New Instrument*. In this treatise, Bacon called on scientists to adopt a new, empirical approach to science. Although he did not discount logic, hypotheses, and mathematics entirely, Bacon emphasized the role of the experimental method in science. Bacon's *The New Instrument* provided a sort of road map for the new order of scientists in Europe who began to value experiments and data over logic and hypotheses. Bacon began by criticizing the scholastics who propagated the reverence for the ancient authorities and thus stifled new learning:

> Those who have taken upon them to lay down the law of nature as a thing already searched out and understood, whether they have spoken in simple assurance or professional affectation, have therein done philosophy and the sciences great injury. For as they have been successful in inducing belief, so they have been effective in quenching and stopping inquiry; and have done more harm by spoiling and putting an end to other men's efforts than good by their own.

Later, Bacon continued his critique of the overdependence on logic in science:

> As the sciences which we now have do not help us in finding out new works, so neither does the logic which we now have help us in finding out new sciences.
>
> The logic now in use serves rather to fix and give stability to the errors which have their foundation in commonly received notions than to help the search after truth. So it does more harm than good.

Bacon proceeded to attack a particular application of logic — one long tied to Aristotelian teaching: "The syllogism is not applied to the first principles of sciences, and is applied in vain to intermediate axioms, being no match for the subtlety of nature. It commands assent therefore to the proposition, but does not take hold of the thing." Bacon's method, on the other hand, was one of induction and open-minded exploration:

> Now my method, though hard to practice, is easy to explain; and it is this. I propose to establish progressive stages of certainty. The evidence of the sense, helped and guarded by a certain process of correction, I retain. But the mental operation which follows the act of sense I for the most part reject; and instead of it I open and lay out a new and certain path for the mind to proceed in, starting directly from the simple sensuous perception.

And Bacon compared the two approaches to knowledge, with one being the clear winner:

> There are and can be only two ways of searching into and discovering truth. The one flies from the senses and particulars to the most general axioms, and from these principles, the truth of which it takes for settled and immovable, proceeds to judgment and to the discovery of middle axioms. And this way is now in fashion. The other derives axioms from the senses and particulars, rising by a gradual and unbroken ascent, so that it arrives at the most general axioms last of all. This is the true way, but as yet untried.
>
> Both ways set out from the senses and particulars, and rest in the highest generalities; but the difference between them is infinite. For the one just glances at experiment and particulars in passing, the other dwells duly and orderly among them.
>
> The one, again, begins at once by establishing certain abstract and useless generalities, the other rises by gradual steps to that which is prior and better known in the order of nature.

Bacon continued throughout his text to give direction to anyone who sought to gain knowledge through empirical study rather than relying on the teachings and logical arguments of the scholastics.

While Francis Bacon's philosophy of science would have a far-reaching influence on new generations of scientists, he never made serious contributions to science itself. Two of his contemporaries, William Gilbert and William Harvey, however, not only adopted the philosophy of the experimental method, but also applied it to particular fields of study to show how experimentation could solve ancient problems in science.

William Gilbert

Rubbing garlic on a magnet will counteract its secret powers. Similarly, diamonds will disrupt the magnet's attractive and repulsive properties. On the other hand, dipping a magnet into a vat filled with goat's blood will increase its power. A magnet may be used to reveal an adulterous woman or to cure a speech defect. Silly superstitions? Of course they are. Yet, when William Gilbert became interested in the mysterious powers of the loadstone (lodestone), or natural magnet, these superstitions — and many others like them — were ingrained in the minds of the educated and uneducated alike. Gilbert not only dispelled these myths, he also discovered many important properties of magnetism. And in the process, he laid the foundation for perhaps the most important characteristic of modern science — the experimental method.

William Gilbert (1544–1603) was born into relatively comfortable surroundings. He attended St. John's College, Cambridge, where he eventually studied medicine. Gilbert established a successful medical practice in London and became a member — and eventual president — of the highly regarded Royal College of Physicians. His reputation grew to the point that Gilbert was appointed personal physician to Queen Elizabeth I, a position from which he was able to carry out his magnetic experiments with the queen's blessing and support. Gilbert never married, seemingly content to spend his life exploring the mysteries of nature and magnetic power. He died, probably of the bubonic plague, while still in service to Queen Elizabeth.

William Gilbert was an early Copernican, calling Copernicus "the restorer of astronomy." In his monumental study in experimental science, *On the Magnet and Magnetic Bodies, and on the Great Magnet the Earth*, Gilbert postulated that the power behind the earth's daily motion was magnetic: "Diurnal motion [of the earth] is due to causes which have now to be sought, arising from magnetick vigour and from the confederated bodies; that is to say why the diurnal rotation of the earth is completed in the space of twenty-four hours."

This explicit acceptance of one of the main tenants of Copernicanism, a moving earth, occurred at a dangerous time for Copernicans in Europe. In only a few years, Galileo became embroiled with the church over the movement of the earth, and in the same year that *On the Magnet* was published (1600), Giordano Bruno was burned at the stake by the Inquisition for heresy. Bruno's condemnation arose from his radical views going far beyond Copernicanism. His suggestion that the stars were suns themselves, and the implication of an infinite universe with other planets and even other living creatures, threatened many of the church's central teachings. Although Gilbert was less radical in his beliefs, and Elizabethan England was a relatively tolerant place, his wholehearted adoption of Copernicanism at the very least placed him outside the mainstream of political, philosophical, and religious doctrine.

Gilbert, like Copernicus, Galileo, Kepler, and so many others trying to break the shackles of traditional natural philosophy, held little respect for the professors and academicians who clung to ancient beliefs in science. In the preface to *On the Magnet*, Gilbert addressed those who he considered the enemies of the new philosophy:

> But why should I, in so vast an Ocean of Books by which the minds of studious men are troubled and fatigued, through which very foolish productions of the world and unreasoning men are intoxicated, and puffed up, rave and create literary broils, and while professing to be philosophers, physicians, mathematicians and astrologers, neglect and despise men of learning: why should I, I say, add aught further to this so-perturbed republick of letters, and expose this noble philosophy, which seems new and incredible by reason of so many things hitherto unrevealed, to be damned and torn to pieces by the maledictions of those who are either already sworn to the opinions of other men, or are foolish corrupters of good arts, learned idiots, grammatists, sophists, wranglers, and perverse little folk? But to you alone, true philosophizers, honest men who seek knowledge not from books only but from things themselves, have I addressed these magnetical principles in this new sort of Philosophizing.

Gilbert's contempt for the "learned idiots" is evident throughout *On the Magnet*, as he presented a "new sort of Philosophizing," namely the experimental method.

On the Magnet is not the first book on the subject of magnetism. This mysterious action, present in nature in the form of the loadstone, was of interest to many, from the natural philosophers who studied it to the sailors who depended on it (in the form of a compass) to find their way on the sea. There is evidence that ancient cultures knew of the powers of loadstone, and that medieval Chinese sailors were using the naturally occurring mineral in primitive compasses. The eventual appearance in Europe of the magnetic compass proved invaluable to seagoing nations. The question on many minds, however, was: How does the compass work? Does it point north because it is somehow attracted by Polaris, the North Star? Some postulated that there existed some large magnetic island in the far north that attracted the needle of the compass.

Gilbert believed he had found the answer to the mystery of the compass. The earth, he claimed, acts as a large magnet! Now, this was not simply a wild theory: Gilbert performed elaborate experiments to show how such a phenomenon could occur. He constructed a spherical loadstone that he called the terrella, or little earth, and proceeded to conduct experiments with magnetized needles and bars. Gilbert took his theory a step further when he concluded that the earth's rotation is due to its magnetic character, confirming in his mind the Copernican theory. His ideas on the magnetic earth were enthusiastically adopted by many of his contemporaries, including Kepler, who postulated that the magnetic force of the earth also existed in the other planets, as well as the sun — a simple and elegant (if ultimately incorrect) mechanism for planetary orbits.

Gilbert introduced many important theories, ideas, and terms in *On the Magnet*. He was the first to conceive of the laws of attraction and repulsion and the first to use the terms north and south poles for magnets; he differentiated between electric and magnetic attraction and even coined the term electricity, from the Greek word for amber.

As important as Gilbert's particular discoveries were to the sciences of magnetism and electricity, perhaps his most important contribution to science was his method. William Gilbert is often called the father of the experimental method. In *On the Magnet*, Gilbert investigated the properties of magnetism and static electricity, often using ingenious devices and apparatuses. His experiments showed that a magnet could be created by rubbing other metals with a magnet. Gilbert demonstrated that magnets cut in half formed new magnets complete with north and south poles. He also proved that high temperatures could destroy the power of a magnet. Gilbert even invented the first device that measured electricity — the electroscope. Recalling that *On the Magnet* preceded the published works of both Galileo and Kepler confirms Gilbert's place in the history of science.

In the preface to *On the Magnet*, Gilbert clearly outlined his plan of experimentation and observation:

> Clearer proofs, in the discovery of secrets, and in the investigation of the hidden causes of things, being afforded by trustworthy experiments and demonstrated arguments, than by the probably guesses and opinions of the ordinary professors of philosophy: so, therefore, that the noble substance of that great magnet, our common mother (the earth), hitherto quite unknown, and the conspicuous and exalted powers of this our globe, may be better understood, we have proposed to begin with the common magnetick, stony, and iron material, and with magnetical bodies, and with the nearer parts of the earth which we can reach with our hands and perceive with our senses; and then to proceed with demonstrable magnetick experiments; and to penetrate, for the first time, into the innermost parts of the earth.

Gilbert set out to present an experimental method, something no one before him had so consciously attempted. In fact, Gilbert supported one of the primary tenants of experimental science, repeatability of the experiments:

> Whoso desireth to make trial of the same experiments, let him handle the substances, not negligently and carelessly, but prudently, deftly, and in the proper way; nor let him (when a thing doth not succeed) ignorantly denounce our discoveries: for nothing hath been set down in these books which hath not been explored and many times performed and repeated amongst us.

And finally, Gilbert paid homage to the revered Greek philosophers, but not in the same way that the scholastics blindly followed their authority:

> Wherefore we but seldom quote ancient Greek authors in our support, because neither by using Greek arguments or Greek words can the truth be demonstrated or elucidated either more precisely or more significantly.... To those early forefathers of philosophy, Aristotle, Theophrastus, Ptolemy, Hippocrates, and Galen, let due honor be ever paid: for by them wisdom hath been diffused to posterity; but our age hath detected and brought to light very many facts which they, were they now alive, would gladly have accepted.

Gilbert paid Aristotle and the rest the ultimate compliment: they would have had an open mind and "gladly" accept the evidence from his experiments.

In one magnificent passage, Gilbert mockingly summarized some of the myths surrounding the magnet, addressed the scholastics in an equally mocking voice, and continued in more detail with his plan for a new experimental philosophy:

> Or that when weak and dulled the virtue is renewed by goats' blood.... Or that Goats' blood sets a loadstone free from the venom of a diamond, so that the lost power is revived when bathed in goats' blood by reason of the discord between that blood and the diamond.... Or

that it removed sorcery from women, and put to flight demons.... Or that it has the power to reconcile husbands to their wives, or to recall brides to their husbands.... Or that in a loadstone pickled in the salt of a sucking fish there is power to pick up gold which has fallen into the deepest wells.... With such idle tales and trumpery do plebian philosophers delight themselves and satiate readers greedy for hidden things, and unlearned devourers of absurdities: But after the magnetick nature shall have been disclosed by the discourse that is to follow and perfected by our labours and experiments, then will the hidden and abstruse causes of so great an effect stand out, sure, proven, displayed, and demonstrated; and at the same time all darkness will disappear, and all error will be torn up by the roots and will lie unheeded; and the foundations of a great magnetick philosophy which have been laid will appear anew, so that high intellects may be no further mocked by idle opinions.

So, we see that William Gilbert's new philosophy was an experimental philosophy. *On the Magnet* represents the first example of the modern experimental method.

But is William Gilbert the first modern scientist? As we have already seen with Kepler and others, no matter how much we hope to use this tag modern to describe the work and philosophy of the leaders of the scientific revolution, these natural philosophers were products of the Renaissance and thus ingrained with Renaissance mysticism and naturalism. Gilbert was no different. He believed that magnetism was caused by an animistic soul. In fact, the title of one of the chapters of *On the Magnet* is "Magnetick force is animate, or imitates life; and in many things surpasses human life, while this is bound up in the organick body." In this chapter, Gilbert wrote: "Loadstone is a wonderful thing in very many experiments, and like a living creature. And one of its remarkable virtues is that which the ancients considered to be a living soul in the sky, in the globes and in the stars, in the sun and in the moon."

Furthermore, Gilbert pointed out that in the Aristotelian cosmology the entire universe is animated except the earth, which lies dead and unmoving at its center. A confirmed Copernican, Gilbert argued that the earth moves, along with the other planets. And what is the force behind this movement? "We, however, consider that the whole universe is animated, and that all the globes, all the stars, and also the noble earth have been governed since the beginning by their own appointed souls."

Then, in spite of his pledge to quote Greek authorities sparingly, Gilbert did just that in order to support his animistic theories: "Wherefore Thales, not without cause (as Aristotle relates in his book *De Anima*), held that the loadstone was animate, being a part and a choice offspring of its animate mother the earth."

It should not come as a surprise that Gilbert harbored such philosophies. Animism, mysticism, and magic were integral components of Renaissance philosophy. Added to this tradition is the fact that force at a distance was very difficult to comprehend. A magnet attracting iron particles makes the particles jump across empty space for no apparent reason. The same difficulty stumped Isaac Newton almost a century later as he tried to understand how gravitational attraction works when there is apparently only empty space between the heavenly bodies. Gilbert was simply proposing a mechanism to explain magnetic attraction and repulsion using the only philosophical tools he had at his disposal. With this in mind, we can understand what historian Paoli Rossi meant when he asked the question, was *On the Magnet* "the last example of Renaissance naturalism or the first of modern experimental science?" Perhaps the answer is yes to both.

William Harvey

William Harvey (1578–1657) was the son of a public official whose position afforded the family a comfortable lifestyle. As a boy, Harvey received a classical education before matriculating to Cambridge. Upon graduation, he traveled to Italy to study medicine at the University of Padua. This proved to be a most fortuitous choice, as Harvey's teacher was the great anatomist Hieronymus Fabricius. After obtaining his medical degree at Padua, Harvey returned home to England. He was soon awarded another medical degree from Cambridge and elected to the College of Physicians. Shortly thereafter, Harvey was placed in charge of St. Bartholomew's Hospital in London, where he spent a good deal of his adult life. In addition to his duties at the hospital, Harvey maintained a thriving private practice and became physician to King James I and later his son, Charles I. When civil war broke out in 1640, Harvey remained loyal to the crown and fled with Charles to Oxford. Harvey eventually retired from his several positions to live a quiet life until his death in 1657.

William Harvey began his medical studies in Padua at a time when the study of medicine meant the study of authorities such as Hippocrates and Galen. Galen was a second-century Roman physician and philosopher whose teachings had been studied and closely followed for centuries. In Galen's time, Roman law forbade dissection of human cadavers, so Galen studied the anatomy of monkeys, a fact that seemed lost in antiquity until Harvey began pointing out mistakes in Galen's conclusions. Galen taught, for instance, that blood was continuously produced by the liver and moved through the body to the various organs to be used as a sort of fuel.

Even though Galen was convinced of the importance of dissection and observation (albeit of monkeys due to the Roman law), through the centuries physicians became less and less interested in this sort of hands on study and relied instead on the words of Galen himself. This began to change in the sixteenth century with the work of the Belgian physician Andreas Vesalius. Vesalius studied human anatomy by performing dissections himself, and insisted that his students do the same. He published his very influential treatise, *On the Workings of the Human Body*, in 1543 — the same year that Copernicus' great work, *On the Revolutions*, appeared. In the dedication to *On the Workings of the Human Body*, Vesalius lamented the state of medical education in the universities:

> And as everything is being thus wrongly taught in the universities and as days pass in silly questions, fewer things are placed before the spectators in all that confusion than a butcher in a market could teach a doctor. I pass over any number of schools where dissecting the structure of the human body is scarcely ever considered; so far has the ancient art of medicine fallen from its early glory many years past.

Furthermore, Vesalius wrote, blind adherence to the authority of Galen was a stain on the study of medicine:

> To this man they have all so entrusted their faith that no doctor has been found who believes he has ever discovered even the slightest error in all the anatomical volumes of Galen, much less that such a discovery is possible: even though (notwithstanding that Galen often corrects himself, that more than once after learning better he points out in some books a careless error he has made in others, and that he often contradicts himself) — even though it is just

now known to us from the reborn art of dissection, from the careful reading of Galen's books, and from the welcome restoration of many portions thereof, that he himself never dissected a human body, but in fact was deceived by his monkeys.

The resurrection of dissection and observation in medical training had a direct impact on William Harvey. Harvey studied under Fabricius at Padua. Fabricius had previously studied under Girolamo Fabrito, and Fabrito's teacher was none other than Vesalius himself. So, we can see that Harvey was a direct intellectual descendent of Vesalius, and this lineage would serve him well in his life's work. In Harvey's classic *On the motion of the heart and blood*, published in 1628, he confirmed this belief in dissection and observation — essentially the experimental method for anatomy:

> True philosophers, who are only eager for truth and knowledge, never regard themselves as already so thoroughly informed, but that they welcome further information from whomsoever and from wheresoever it may come; nor are they so narrow-minded as to imagine any of the arts or sciences transmitted to us by the ancients, in such a state of forwardness or completeness, that nothing is left for the ingenuity and industry of others.... I profess both to learn and to teach anatomy, not from books but from dissections; not from the positions of philosophers but from the fabric of nature.

Harvey, in carrying out this program of dissections and observations, carved out a place for himself alongside William Gilbert as one of the founders of the experimental philosophy.

When William Harvey began his anatomical explorations, physicians held many misconceptions about the role of the heart and blood in the human body. Harvey acknowledged these long-held beliefs, most of which descended from Galen:

> As we are about to discuss the motion, action, and use of the heart and arteries, it is imperative on us first to state what has been thought of these things by others in their writings, and what has been held by the vulgar and by tradition, in order that what is true may be confirmed, and what is false set right by dissection, multiplied experience, and accurate observation.

Harvey discussed the common thinking that blood was manufactured by the liver and consumed by the various organs. Other misconceptions concerning the role of the veins and arteries, as well as the construction and actions of the heart itself, set Harvey on a path of dissection, vivisection, and experimentation that eventually led to one of the most significant discoveries in the history of science:

> Since, therefore, from the foregoing considerations and many others to the same effect, it is plain that what has heretofore been said concerning the motion and function of the heart and arteries must appear obscure, inconsistent, or even impossible to him who carefully considers the entire subject, it would be proper to look more narrowly into the matter to contemplate the motion of the heart and arteries, not only in man, but in all animals that have hearts; and also, by frequent appeals to vivisection, and much ocular inspection, to investigate and discern the truth.... The motion of the heart, then, is entirely of this description, and the one action of the heart is the transmission of the blood and its distribution, by means of the arteries, to the very extremities of the body; so that the pulse which we feel in the arteries is nothing more than the impulse of the blood derived from the heart.

Harvey's conclusion, that the blood was circulated throughout the body, was revolutionary and, not surprisingly, widely criticized by his fellow physicians. Harvey, like

Copernicus before him and also like contemporaries such as Galileo and Kepler, met with strong opposition. And Harvey, like Copernicus, Galileo, and Kepler, addressed these opponents in his publication:

> These views as usual, pleased some more, others less; some chid and calumniated me, and laid it to me as a crime that I had dared to depart from the precepts and opinions of all anatomists; others desired further explanations of the novelties, which they said were both worthy of consideration, and might perchance be found of signal use. At length, yielding to the requests of my friends, that all might be made participators in my labors, and partly moved by the envy of others, who, receiving my views with uncandid minds and understanding them indifferently, have essayed to traduce me publicly, I have moved to commit these things to the press, in order that all may be enabled to form an opinion both of me and my labours. This step I take all the more willingly, seeing that Hieronymus Fabricius of Aquapendente, although he has accurately and learnedly delineated almost every one of the several parts of animals in a special work, has left the heart alone untouched. Finally, if any use or benefit to this department of the republic of letters should accrue from my labours, it will, perhaps, be allowed that I have not lived idly....

William Harvey's importance to the history of science — like that of William Gilbert's — lies as much in his methods as in his conclusions. Harvey, like Gilbert, relied on the experimental philosophy to support his conclusions. In *On the motion*, Harvey described experiment after experiment — dissections, vivisections, etc. — that support his conclusion that the blood must circulate in the bodies of animals. One of these was a simple observation which, Harvey argued, made Galen's theory of the blood being produced by the liver untenable:

> Let us assume, either arbitrarily or from experiment, the quantity of blood which the left ventricle of the heart will contain when distended, to be, say, two ounces, three ounces, or one ounce and a half — in the dead body I have found it to hold upwards of two ounces. Let us assume further how much less the heart will hold in the contracted than in the dilated state; and how much blood it will project into the aorta upon each contraction; and all the world allows that with the systole something is always projected, a necessary consequence demonstrated in the third chapter, and obvious from the structure of the valves; and let us suppose as approaching the truth that the fourth, or fifth, or sixth, or even but the eighth part of its charge is thrown into the artery at each contraction; this would give either half an ounce, or three drachms, or one drachm of blood as propelled by the heart at each pulse into the aorta; which quantity, by reason of the valves at the root of the vessel, can by no means return into the ventricle. Now, in the course of half an hour, the heart will have made more than one thousand beats, in some as many as two, three, and even four thousand. Multiplying the number of drachms propelled by the number of pulses, we shall have either one thousand half ounces, or one thousand times three drachms, or a like proportional quantity of blood, according to the amount which we assume as propelled with each stroke of the heart, sent from this organ into the artery — a larger quantity in every case than is contained in the whole body!

Harvey has shown, with simple arithmetic and using very conservative estimates, that the amount of blood flowing through the blood vessels in half an hour is more than the entire body contains! Can this much blood be continually produced and consumed in the body over an extended period of time? Harvey thought it very unlikely:

> But let it be said that this does not take place in half an hour, but in an hour, or even in a day; any way, it is still manifest that more blood passes through the heart in consequence of

its action, than can either be supplied by the whole of the ingesta, or than can be contained in the veins at the same moment.

So we see that Harvey concluded that to assume that blood is continuously produced and consumed in the body is essentially a physical impossibility. The conclusion, then, is that the heart recirculates the same blood throughout the body.

Perhaps Harvey's most conclusive experiment — and one of the most famous experiments in the history of science — involved a careful observation of the movement of blood through a human arm after the application of a ligature, or tourniquet. Harvey described in great detail his experiments:

> Now let anyone make an experiment upon the arm of a man, either using such a fillet as is employed in blood-letting, or grasping the limb lightly with his hand, the best subject for it being one who is lean, and who has large veins, and the best time after exercise, when the body is warm, the pulse is full, and the blood carried in larger quantity to the extremities, for all then is more conspicuous; under such circumstances let a ligature be thrown about the extremity, and drawn as tightly as can be borne, it will first be perceived that beyond the ligature, neither in the wrist nor anywhere else, do the arteries pulsate, at the same time that immediately above the ligature the artery begins to rise higher at each diastole, to throb more violently, and to swell in its vicinity with a kind of tide, as if it strove to break through and overcome the obstacle to its current; the artery here, in short, appears as if it were preternaturally full. The hand under such circumstances retains its natural colour and appearance; in the course of time it begins to fall somewhat in temperature, indeed, but nothing is drawn into it.... After the bandage has been kept on for some short time in this way, let it be slackened a little, brought to that state or term of medium tightness which is used in bleeding, and it will be seen that the whole hand and arm will instantly become deeply coloured and distended, and the veins show themselves tumid and knotted; after ten or twelve pulses of the artery, the hand will be perceived excessively distended, injected, gorged with blood, drawn, as it is said, by this medium ligature, without pain, or heat, or any horror of a vacuum, or any other cause yet indicated.... If the finger be applied over the artery as it is pulsating by the edge of the fillet, at the moment of slackening it, the blood will be felt to glide through, as it were, underneath the finger; and he, too, upon whose arm the experiment is made, when the ligature is slackened, is distinctly conscious of a sensation of warmth, and of something, viz., a stream of blood suddenly making its way along the course of the vessels and diffusing itself through the hand, which at the same time begins to feel hot, and becomes distended.... As we had noted, in connexion with the tight ligature, that the artery above the bandage was distended and pulsated, not below it, so, in the case of the moderately tight bandage, on the contrary, do we find that the veins below, never above, the fillet, swell, and become dilated, whilst the arteries shrink; and such is the degree of distension of the veins here, that it is only very strong pressure that will force the blood beyond the fillet, and cause any of the veins in the upper part of the arm to rise.

What is the explanation of these curious results, Harvey asked? For him, the answer was obvious. When the ligature was tight, the flow of blood was cut off in both the arteries and the veins. The lower arm, cut off from the blood supply, became pale and cold. The upper arm on the other hand was warm and throbbing with blood unable to pass the ligature. When Harvey loosened the ligature a little, it allowed blood to flow through the arteries into the lower arm, the arteries being deeper in the body than the veins and therefore not restricted by the loosened ligature. Harvey noted that with the blood now flowing into the lower arm it became warm again, even swollen, as blood rushed into the lower arm but could not return through the veins closer to the skin surface. So, Harvey concluded:

> From these facts it is easy for every careful observer to learn that the blood enters an extremity by the arteries; for when they are effectually compressed nothing is drawn to the member; the hand preserves its colour; nothing flows into it, neither is it distended; but when the pressure is diminished, as it is with the bleeding fillet, it is manifest that the blood is instantly thrown in with force, for then the hand begins to swell; which is as much as to say, that when the arteries pulsate the blood is flowing through them, as it is when the moderately tight ligature is applied; but where they do not pulsate, as, when a tight ligature is used, they cease from transmitting anything, they are only distended above the part where the ligature is applied. The veins again being compressed, nothing can flow through them; the certain indication of which is, that below the ligature they are much more tumid than above it, and than they usually appear when there is no bandage upon the arm.... It therefore plainly appears that the ligature prevents the return of the blood through the veins to the parts above it, and maintains those beneath it in a state of permanent distension. But the arteries, in spite of its pressure, and under the force and impulse of the heart, send on the blood from the internal parts of the body to the parts beyond the ligature. And herein consists the difference between the tight and the medium ligature, that the former not only prevents the passage of the blood in the veins, but in the arteries also; the latter, however, whilst it does not prevent the force of the pulse from extending beyond it, and so propelling the blood to the extremities of the body, compresses the veins, and greatly or altogether impedes the return of the blood through them.... Seeing, therefore, that the moderately tight ligature renders the veins turgid and distended, and the whole hand full of blood, I ask, whence is this? Does the blood accumulate below the ligature coming through the veins, or through the arteries, or passing by certain hidden porosities? Through the veins it cannot come; still less can it come through invisible channels; it must needs, then, arrive by the arteries in conformity with all that has been already said. That it cannot flow in by the veins appears plainly enough from the fact that the blood cannot be forced towards the heart unless the ligature be removed; when this is done suddenly all the veins collapse, and disgorge themselves of their contents into the superior parts, the hand at the same time resumes its natural pale colour, the tumefaction and the stagnating blood having disappeared.... Moreover, he whose arm or wrist has thus been bound for some little time with the medium bandage, so that it has not only got swollen and livid but cold, when the fillet is undone is aware of something cold making its way upwards along with the returning blood, and reaching the elbow or the axilla. And I have myself been inclined to think that this cold blood rising upwards to the heart was the cause of the fainting that often occurs after blood-letting: fainting frequently supervenes even in robust subjects, and mostly at the moment of undoing the fillet, as the vulgar say, from the turning of the blood.

Harvey's critics, even before he published *On the motion*, were quick to point out that there was no apparent avenue for the blood to pass from the arteries to the veins, making his theory very suspect. Harvey himself did not know the answer to this criticism, but he did speculate:

> Farther, when we see the veins below the ligature instantly swell up and become gorged, when from extreme tightness it is somewhat relaxed, the arteries meantime continuing unaffected, this is an obvious indication that the blood passes from the arteries into the veins, and not from the veins into the arteries, and that there is either an anastomosis of the two orders of vessels, or porosities in the flesh and solid parts generally that are permeable to the blood. It is farther an indication that the veins have frequent communications with one another,

because they all become turgid together, whilst under the medium ligature applied above the elbow; and if any single small vein be pricked with a lancet, they all speedily shrink, and disburthening themselves into this they subside almost simultaneously.

Harvey's failure to correctly ascertain how blood was transferred from the arteries to the veins was understandable. The capillaries are microscopic, and when Harvey performed his experiments and made his observations the modern microscope had yet to be invented. It would be a generation later before an Italian physician named Marcello Malpighi observed capillaries under the microscope and postulated that they were the pathways by which this transfer of blood occurred. Malpighi and other early microscopists, like Antonie van Leeuwenhoek and Robert Hooke, used the new instrument to discover an unseen world here on earth as Galileo had used the telescope to discover unseen worlds outside of earth. In his classic work, *Micrographia*, Hooke expressed this sentiment:

> The next care to be taken, in respect of the Senses, is a supplying of their infirmities with Instruments, and, as it were, the adding of artificial Organs to the natural; this in one of them has been of late years accomplisht with prodigious benefit to all sorts of useful knowledge, by the invention of Optical Glasses. By the means of Telescopes, there is nothing so far distant but may be represented to our view; and by the help of Microscopes, there is nothing so small, as to escape our inquiry; hence there is a new visible World discovered to the understanding. By this means the Heavens are open'd, and a vast number of new Stars, and new Motions, and new Productions appear in them, to which all the ancient Astronomers were utterly Strangers. By this the Earth it self, which lyes so neer us, under our feet, shews quite a new thing to us, and in every little particle of its matter; we now behold almost as great a variety of Creatures, as we were able before to reckon up in the whole Universe it self.

Harvey performed another simple experiment to ascertain the function of the valves in veins. Harvey's teacher, Fabricius, is credited with discovering these valves, although he did not understand their function. Harvey observed that a few simple manipulations to the veins in the arm (or anywhere else in the body) revealed the true nature of these valves:

> And now if you press the blood from the space above one of the valves ... and keep the point of a finger upon the vein inferiorly, you will see no influx of blood from above; The blood being thus pressed out and the vein emptied, if you now apply a finger of the other hand upon the distended part of the vein above the valve ... and press downwards, you will find that you cannot force the blood through or beyond the valve; but the greater effort you use, you will only see the portion of vein that is between the finger and the valve become more distended, that portion of the vein which is below the valve remaining all the while empty.... It would therefore appear that the function of the valves in the veins is the same as that of the three sigmoid valves which we find at the commencement of the aorta and pulmonary artery, viz., to prevent all reflux of the blood that is passing over them.

In other words, the valves allowed a one way flow of the blood through the veins to the heart.

Harvey continued on in this way, further demonstrating the nature of the flow of the blood through the veins and the role of the valves. This, he claimed, is further evidence of the circulation of the blood:

> And now compute the quantity of blood which you have thus pressed up beyond the valve, and then multiplying the assumed quantity by one thousand, you will find that so much

blood has passed through a certain portion of the vessel; and I do now believe that you will find yourself convinced of the circulation of the blood, and of its rapid motion.

Harvey then made a remarkable claim. In generations to come, as the experimental method developed in science, one of the standards upon which it was built was repeatability. Harvey became one of the first natural philosophers to allude to such a standard when he invited the readers to perform the same experiments and draw their own conclusions:

> But if in this experiment you say that a violence is done to nature, I do not doubt but that, if you proceed in the same way, only taking as great a length of vein as possible, and merely remark with what rapidity the blood flows upwards, and fills the vessel from below, you will come to the same conclusion.

In addition to his role as an early proponent of the experimental philosophy, Harvey is also attributed with a key contribution to the burgeoning mechanical philosophy. Harvey made analogies to machines and mechanical devices when explaining the role of the heart in the circulation of the blood. For instance, when Harvey wrote about the motion of the heart itself, he pointed out:

> These two motions, one of the ventricles, the other of the auricles, take place consecutively, but in such a manner that there is a kind of harmony or rhythm preserved between them, the two concurring in such wise that but one motion is apparent, especially in the warmer blooded animals, in which the movements in question are rapid. Nor is this for any other reason than it is in a piece of machinery, in which, though one wheel gives motion to another, yet all the wheels seem to move simultaneously; or in that mechanical contrivance which is adapted to firearms, where, the trigger being touched, down comes the flint, strikes against the steel, elicits a spark, which falling among the powder, ignites it, when the flame extends, enters the barrel, causes the explosion, propels the ball, and the mark is attained — all of which incidents, by reason of the celerity with which they happen, seem to take place in the twinkling of an eye.

Although Harvey's philosophy was more Aristotelian and Renaissance vitalism than mechanical philosophy, musings on the mechanical nature of the heart influenced René Descartes and the new ideas in mechanical philosophy. The idea of the heart acting as a mechanical pump inspired Descartes to explore the possibility of a universe and everything in it, behaving like a machine (see Chapter 5).

Harvey attached a deep significance to the idea that the blood moved through the body in a circular path, the circle being both the ancient and the Renaissance ideal configuration:

> I began to think whether there might not be a Motion, As It Were, In A Circle. Now, this I afterwards found to be true; and I finally saw that the blood, forced by the action of the left ventricle into the arteries, was distributed to the body at large, and its several parts, in the same manner as it is sent through the lungs, impelled by the right ventricle into the pulmonary artery, and that it then passed through the veins and along the vena cava, and so round to the left ventricle in the manner already indicated. This motion we may be allowed to call circular, in the same way as Aristotle says that the air and the rain emulate the circular motion of the superior bodies; for the moist earth, warmed by the sun, evaporates; the vapours drawn upwards are condensed, and descending in the form of rain, moisten the earth again. By this arrangement are generations of living things produced; and in like manner are tempests and meteors engendered by the circular motion, and by the approach and recession of the sun.

Harvey continued to make references to the vitalistic and spiritual nature of the blood, and attached significance to the centrality of the heart:

> And similarly does it come to pass in the body, through the motion of the blood, that the various parts are nourished, cherished, quickened by the warmer, more perfect, vaporous, spirituous, and, as I may say, alimentive blood; which, on the other hand, owing to its contact with these parts, becomes cooled, coagulated, and so to speak effete. It then returns to its sovereign, the heart, as if to its source, or to the inmost home of the body, there to recover its state of excellence or perfection. Here it renews its fluidity, natural heat, and becomes powerful, fervid, a kind of treasury of life, and impregnated with spirits, it might be said with balsam. Thence it is again dispersed. All this depends on the motion and action of the heart.

Furthermore, Harvey continued, the heart is to the body as the sun is to the universe; the heart is the "sun of the microcosm," in Harvey's poetic words:

> The heart, consequently, is the beginning of life; the sun of the microcosm, even as the sun in his turn might well be designated the heart of the world; for it is the heart by whose virtue and pulse the blood is moved, perfected, and made nutrient, and is preserved from corruption and coagulation; it is the household divinity which, discharging its function, nourishes, cherishes, quickens the whole body, and is indeed the foundation of life, the source of all action.

We can see from these passages, and many others just like them, that although it is tempting to see Harvey as a modern experimental scientist, he was equally a product of the Renaissance naturalism and Aristotelianism on which he was educated.

In addition to his other contributions to modern science, Harvey also worked in what today would be called embryology. Through many experiments, especially with the eggs of chickens, Harvey embraced the theory of epigenesis, a term that he coined but whose principles were laid out centuries before by Aristotle. The theory of epigenesis claims that life in an egg grows from a homogeneous mass into a diverse body. The theory's primary rival was called preformation. Preformationists claimed that life began as a perfect miniature of its final form. Even when microscopists like van Leeuwenhoek (who was the first to see sperm under a microscope) began studying eggs, the form of the embryo was unclear. Harvey postulated that mammals reproduced through the contact of sperm and egg, several hundred years before the mammalian egg was observed under the microscope.

Natural philosophy, or science, in Europe at the beginning of the seventeenth century was dominated by the Aristotelian deductive philosophy. The empirical philosophy and the experimental method were largely ignored. Thanks to the influential writings of Francis Bacon, and the breakthrough discoveries of William Gilbert and William Harvey, by mid-century the experimental method was being discussed and debated by philosophers all over the continent. With the new discoveries of Gilbert and Harvey, experimentation was beginning to find its place in science.

PRIMARY SOURCES

Gilbert, William. *On the Magnet*, translated by Silvanus Thompson. New York: Basic Books, 1958.
Harvey, William. *On the Motion of the Heart and Blood in Animals*, translated by Robert Willis. *Scientific Papers; Physiology, Medicine, Surgery, Geology, with Introductions, Notes and Illustrations*, the Harvard Classics, vol. 38. New York: P. F. Collier & Son, 1910.

Hooke, Robert. *Micrographia*. Project Gutenberg. http://www.gutenberg.org/files/15491/15491-h/15491-h.htm.

Vesalius, Andreas. *On the Fabric of the Human Body*, annotated translation by Daniel Garrison and Malcolm Hast. Evanston, IL: Northwestern University, 2003.

OTHER SOURCES

Bylebyl, Jerome. *William Harvey and His Age: The Professional and Social Context of the Discovery of the Circulation*. Baltimore: Johns Hopkins University Press, 1979.

French, R. K. *William Harvey's Natural Philosophy*. New York: Cambridge University Press, 1994.

Gregory, Andrew. *Harvey's Heart, the Discovery of Blood Circulation*. Cambridge, UK: Icon, 2001.

Kelly, Suzanne. *The De Mundo of William Gilbert*. Amsterdam: Hertzberger, 1965.

King, W. James. *The Natural Philosophy of William Gilbert and His Predecessors*. Washington, DC: Smithsonian Institute, 1959.

Mottelay, P. Fleury. *William Gilbert of Colchester*. New York: Wiley and Sons, 1893.

Neil, Eric. *William Harvey and the Circulation of the Blood*. London: Priory Press, 1975.

O'Malley, C.D. *Andreas Vesalius of Brussels*. Berkeley: University of California Press, 1964.

Roller, Duane Henry DuBose. *The De Magnete of William Gilbert*. Amsterdam: Hertzberger, 1959.

Shackelford, Jole. *William Harvey: And the Mechanics of the Heart*. Oxford: New York: Oxford University Press, 2003.

Whitteridge, Gweneth. *William Harvey and the Circulation of the Blood*. New York: American Elsevier, 1971.

CHAPTER 5

Descartes, Boyle, and the
Mechancial Philosophy

If you would be a real seeker after truth, you must at least once in your life doubt, as far as possible, all things.— René Descartes, *Discourse on Method* (1637)

Thus, the universe was, once framed by God and the laws of motion, settled and everything was upheld by his perpetual concourse and general providence, the same philosophy teaches that the phenomena of the world are physically produced by the mechanical properties of the parts of matter, and that they operate upon one another according to mechanical laws.— Robert Boyle, *On the Excellency and Grounds of the Corpuscular or Mechanical Philosophy* (1674)

Copernicus began a scientific revolution by proposing that the sun lay at the center of the universe and the earth moved through the heavens. His heliocentric theory, however, conflicted with the traditional Aristotelian physics that assumed a heavy, motionless earth at the center of the cosmos. Because of this, the science of motion in the earthly realm languished. Galileo, in addition to his monumental discoveries with the telescope, introduced a new way of thinking about and describing motion on earth. He proposed a new physics to replace the traditional Aristotelian model that had prevailed for centuries. Galileo's physical model was both empirical and mathematical. It did, however, leave questions as to the *cause* of motion.

Into this void stepped the mechanical philosophers. For ages the accepted explanations for motion, heavenly or terrestrial, involved various forms of Aristotelian philosophy often combined with notions of animate forces continually at play. Movement of inanimate objects was caused by the same sort of forces that caused movement in animals and humans. Animistic, occult, and other supernatural explanations prevailed.

The scholastics of the Renaissance universities were entrenched as the ultimate authorities on natural philosophy. During the seventeenth century, original thinkers such as René Descartes and Robert Boyle led a surge of new explanatory structures for motion,

both on earth and in the heavens. The mechanical philosophy that grew out of their spec-
ulations began to replace the supernatural and animistic views with natural, physical, and
mechanical theories. And although much of their strictly mechanical view of the universe
was ultimately discarded, the rejection of the supernatural in favor of the physical by the
mechanical philosophers permanently changed the way scientists approached their work.

René Descartes

The name most universally associated with the mechanical philosophy is that of the
French philosopher and mathematician René Descartes. Descartes (1596–1650) is often
referred to as the "father of modern philosophy." Perhaps the most famous line in the history of philosophy comes from the pen of Descartes: "I think, therefore I am." This line
is a result of Descartes' belief that the essence of humanity — in fact the only unassailable
proof of the existence of humankind, is our capability for rational thought. Descartes
directed his own powers of rational thought towards many subjects in philosophy, but it
will be his ideas on natural philosophy, or science, which will be emphasized here.

Descartes was born into a prominent French family. His father was a member of the
French Parliament and a member of France's political and intellectual elite. His family
was liberally sprinkled with lawyers and physicians, so the best education was a given for
the young René. A sickly child, when Descartes' father sent him to boarding school René
was allowed to sleep late while other boys were awakened early to begin their regimented
day. In school, Descartes studied the traditional subjects of the trivium and the quadrivium, as well as a sprinkling of natural philosophy. The trivium and the quadrivium
together formed the classical seven liberal arts. These disciplines, a mainstay in medieval
and Renaissance universities and dating back to the Pythagorean school of classical Greece,
included grammar, rhetoric, and logic (the trivium), and arithmetic, geometry, astronomy,
and music (the quadrivium). Descartes attended the University of Poitiers, where he studied law, along with philosophy, theology, and medicine.

In spite of the best education available at the time, Descartes would later look back
at his years in school and lament the fact that he had learned nothing from reading from
textbooks and listening to the lectures of the scholastics. In his seminal work, *Discourse
on Method*, Descartes wrote:

> From my childhood ... I was ardently desirous of instruction. But as soon as I had finished
> the entire course of study, at the close of which it is customary to be admitted into the order
> of the learned ... I found myself involved in so many doubts and errors, that I was convinced
> I had advanced no farther in all my attempts at learning, than the discovery at every turn of
> my own ignorance.

So Descartes set his own course for knowledge, a course he was to stay for the remainder
of his life:

> For these reasons, as soon as my age permitted me to pass from under the control of my
> instructors, I entirely abandoned the study of letters, and resolved no longer to seek any
> other science than the knowledge of myself, or of the great book of the world. I spent the
> remainder of my youth in traveling, in visiting courts and armies, in holding intercourse

with men of different dispositions and ranks, in collecting varied experience, in proving myself in the different situations into which fortune threw me, and, above all, in making such reflection on the matter of my experience as to secure my improvement. For it occurred to me that I should find much more truth in the reasonings of each individual with reference to the affairs in which he is personally interested, and the issue of which must presently punish him if he has judged amiss, than in those conducted by a man of letters in his study, regarding speculative matters that are of no practical moment, and followed by no consequences to himself, farther, perhaps, than that they foster his vanity the better the more remote they are from common sense; requiring, as they must in this case, the exercise of greater ingenuity and art to render them probable. In addition, I had always a most earnest desire to know how to distinguish the true from the false, in order that I might be able clearly to discriminate the right path in life, and proceed in it with confidence.

In seventeenth century Europe, authority ruled. The authority of the ancient philosophers, the authority of the university professor, the authority of scripture and the church — all dominated the way Europeans thought and behaved. Descartes revolted against this dogma, vowing to learn only by his own reason and not by anyone's authority: "If you would be a real seeker after truth, you must at least once in your life doubt, as far as possible, all things."

Disenchanted with his education thus far, young Descartes joined the army of the Dutch prince Maurice of Nassau. Here his role was apparently as a gentleman observer and possibly a military engineer — it seems certain that he was nowhere near any real fighting. Later Descartes relates that while in the military he had a series of dreams that set him upon a quest to study science. At about the same time, Descartes met Isaac Beekman, an early proponent of the new mechanical philosophy. Beekman became Descartes' mentor in science and mathematics and inspired the young man to continue his self-education. After his military service ended, Descartes remained in The Netherlands where he spent much of his adult life studying and writing. He somewhat reluctantly moved to Sweden at the urging of Queen Christina to be her tutor and confidant. Unfortunately, the queen, who was an early riser, insisted that the notoriously late sleeping Descartes awake early for their lessons. Unused to the hours and the cold, Descartes succumbed (possibly to pneumonia) and died in 1650.

Descartes wrote and published widely on philosophy, theology, science (natural philosophy), mathematics, and a variety of other subjects. In fact, Cartesian philosophy was an attempt to explain science, religion, and philosophy in one intellectual system. In the end, Descartes created a philosophical system to rival Aristotle. The Cartesian universe was mechanical in nature, yet still based on rational thought and mathematics, not on empirical evidence. Unlike his contemporaries, William Harvey and William Gilbert (and to some extent Galileo), Descartes was not a proponent of experimental science. In fact, Descartes distrusted the senses and thought everyday experience to be an unreliable basis for natural philosophy. In *Discourse on Method*, Descartes explained his own methods in philosophy, theology, and science. He emphasized the importance of one's own ability for rational thought and how one should approach learning and understanding the surrounding world.

Important in its own right, *Discourse on Method* also included three highly influential, even revolutionary, appendices. These appendices were entitled *Optics*, *Meteorology*, and

Geometry. In *Optics*, Descartes reversed the optical pyramid of the ancients, arguing that sight should be understood as light emanating from an object to the eye where it was focused by the lens on to the retina. Descartes also derived the laws of reflection and refraction, the later now known as Snell's Law.

The *Geometry* of Descartes laid the foundation for modern mathematics. In it, Descartes introduced analytic geometry, combining the equations of algebra with the figures of geometry. Today, anytime a graph associated with an algebraic equation is constructed — which will, of course, appear on a grid with horizontal and vertical axes known as the Cartesian coordinate system — credit is due René Descartes and the other independent co–founder of analytic geometry, Pierre Fermat. The (possibly apocryphal) story of the invention of this rectangular coordinate system has the young Descartes lying in his bed observing a fly crawling on the ceiling when it occurs to him that he can describe the fly's position by plotting its distance from each wall. The development of analytic geometry was one stepping stone to other new mathematical disciplines, most importantly calculus.

Although *Discourse on Method* certainly ruffled the feathers of the scholastic elite, it did not contain ideas that were immediately seen as dangerous to the church hierarchy. However, another of Descartes' important works, *The World*, did contain dangerous ideas. In fact, just as he was preparing to publish *The World*, word of Galileo's condemnation by the Inquisition arrived in France and Descartes chose to delay the publication. What was contained in this work that caused Descartes to back away? First of all, Descartes adopted and supported the heliocentric model of Copernicus. This in itself was dangerous, because the Catholic Church had already placed *On the Revolutions* on the list of Index of Prohibited Books. But *The World* also contained an explanation of Descartes' mechanical theories. These theories were questioned by religious authorities because they feared the ideas led to materialism.*

In *The World*, Descartes described nature as mechanical and motion as the direct result of contact between physical objects. Underlying the entire system was the smallest unit of nature, the corpuscle. Descartes wrote:

> Consider that every body can be divided into extremely small parts. I do not wish to determine whether their number is infinite or not; at least it is certain that, with respect to our knowledge, it is indefinite and that we can suppose that there are several millions in the smallest grain of sand our eyes can perceive.

Although sounding much like atomism, Descartes' corpuscle theory had several subtle, and not so subtle, differences. Most importantly, the atomists often argued for the existence of a void, or vacuum, all around us. Descartes, on the other hand, was in firm agreement with Aristotle and argued that a void was impossible. He used an analogy to make his argument:

> For example, if you are placing powder in a jar, you shake the jar and pound against it to make room for more powder. But, if you are pouring some liquid into it, the liquid spontaneously arranges itself in as small a place as one can put it. By the same token, if you consider in this regard some of the experiments the philosophers have been wont to use in showing that there is no void in nature, you will easily recognize that all those spaces that

---

*Here the word materialism is not used in the same sense as it often appears today. Today we think of materialism as synonymous with consumerism — the desire to have more "things." In the seventeenth century, materialism meant an over emphasis on the physical rather than the spiritual. Materialism in philosophy minimized (or did away with entirely) God's role in nature and everyday life.

people think to be empty, and where we feel only air, are at least as full, and as full of the same matter, as those where we sense other bodies.

The debate raged on about the physical, philosophical, and even theological implications of the existence of a vacuum. Atomism, first proposed by Greek philosophers Democritus and Epicurus, was revived by the French natural philosopher Pierre Gassendi. Most atomists argued that the geometry of atoms required that much of space be empty — a vacuum. Descartes and others argued that the void was nothingness which was a philosophical and physical impossibility.

The deciding blow in the debate concerning the existence of a void came when the Italian natural philosopher Evangelista Torricelli invented the barometer. Torricelli's device was a simple tube open at one end, filled with mercury, then turned upside down into a dish filled with more mercury. Torricelli concluded that the space left in the top of the tube after some of the mercury ran out into the dish was empty space — a void. Later experiments confirmed this conclusion. Torricelli also deduced that the column of mercury left in the tube was caused by the atmospheric pressure exerted on the surface of the mercury in the dish — in other words, the column of mercury was a measure of the barometric pressure.

One of the most famous experiments in the history of science employed Torricelli's invention. The French mathematician, philosopher, and inventor Blaise Pascal instructed his brother-in-law to carry two identical mercury barometers up the mountain known as the Puy de Dôme. When his brother-in-law noted that the column of mercury in the tubes fell as the altitude increased, Pascal had experimental proof that atmospheric pressure decreased with altitude.

In spite of mounting evidence, Descartes never accepted the reality of a vacuum, primarily because it conflicted with the mechanical philosophy and his theory of the plenum. Descartes argued that all of space was filled (space was a plenum) and that motion could be understood only as the result of direct contact. When an object moved through space, it replaced the material it moved through and this material in turn filled the space previously occupied by the moving object:

> But you could propose to me here a rather considerable problem, to wit, that the parts composing liquid bodies cannot, it seems, move incessantly, as I have said they do, unless there is some empty space among them, at least in the places from which they depart by virtue of their being in motion. I would have trouble responding to this, had I not recognized through various experiences that all the motions that take place in the world are in some way circular. That is to say, when a body leaves its place, it always enters into that of another, and the latter into that of still another, and so on down to the last which occupies in the same instant the place left open by the first.

Descartes used the analogy of an object moving through water. Although the object is completely surrounded by water, it displaces the water in front of it as it moves and the water in turn immediately fills in the spaces previously occupied by the object:

> Now, when bodies move in the air, we do not usually notice these circular motions, because we are accustomed to conceiving of the air only as an empty space. But look at fish swimming in the pool of a fountain: if they do not approach too near to the surface of the water, they cause great speed. Whence it clearly appears that the water they push before them does not push indifferently all the water of the pool, but only that which can best serve to perfect the circle of the fishes' motion and return to the place they leave behind.

The circular motion Descartes wrote about was central to his mechanical theory. He envisioned a world where movement was caused by vortices spinning in circular motion and continually in contact with matter. He used the theory of vortices to describe all motion. For instance, he explained the motion of the planets as they revolved around the sun by thinking of the sun as the center of a vortex. Much like an object caught in a whirlpool, the planets embedded in these vortices spun around the sun.

In *The World*, Descartes laid out his theory of motion by direct contact in a plenum. In particular, he stated several rules of motion.

> The first is that each individual part of matter always continues to remain in the same state unless collision with others constrains it to change that state. That is to say, if the part has some size, it will never become smaller unless others divide it; if it is round or square, it will never change that shape without others forcing it to do so; if it is stopped in some place, it will never depart from that place unless others chase it away; and if it has once begun to move, it will always continue with an equal force until others stop or retard it.

In this first rule, we can see the seed of the law of inertia. Although Descartes' insistence on treating rest and motion as two different states differs from the modern concept of inertia, his reliance on mechanical and quantitative explanations, rather than occult and animistic ones, greatly influenced the direction of physics.

Descartes' second rule represents a conservation of motion, although Descartes never expressed this motion in terms of momentum. "I suppose as a second rule that, when one of these bodies pushes another, it cannot give the other any motion except by losing as much of its own at the same time; nor can it take away from the other body's motion unless its own is increased by as much."

Lest we forget that Descartes was a devout believer in God's creation of the natural world and his role in establishing the laws which govern nature, Descartes wrote:

> Now it is the case that those two rules manifestly follow from this alone: that God is immutable and that, acting always in the same way, He always produces the same effect. For, supposing that He placed a certain quantity of motions in all matter in general at the first instant He created it, one must either avow that He always conserves as many of them there or not believe that He always acts in the same way.

Although Descartes sought to establish natural mechanisms to explain motion, the reason these natural mechanisms exist in the first place, Descartes explained, is due to God the creator.

Remembered today primarily as a philosopher, Descartes' scientific legacy was of lasting importance. Although his theories on motion have long since been discarded, Cartesian natural philosophy, especially his mechanical philosophy, was the dominant intellectual system of the seventeenth century. He successfully sought to replace the Aristotelian emphasis on the senses with a strictly rational view of physics, particularly through geometry, which Descartes argued was certain knowledge. He, along with the other mechanical philosophers, also challenged the view of a world filled with living principles and replaced it with a mechanical universe that behaved as a machine. His work, along with contemporaries like Robert Boyle, effectively challenged the centuries-old hold of the scholastics on scientific thought and fundamentally changed the way science was understood.

Robert Boyle

Robert Boyle (1627–1691) was a leading proponent of both the experimental and the mechanical philosophy. Both of these movements, so integral to the development of modern science, were in their infancy in the seventeenth century. Boyle conducted careful and repeatable experiments and explained his results in a mechanical language very much different from the prevailing ideas. His mechanical explanations of chemical experiments relied on the basic premises of matter and motion as understood through the emerging corpuscular theory. Like Descartes, Boyle was a life-long and adamant critic of the Aristotelian scholasticism that had so long dominated European thought.

Born into a wealthy and influential Irish family, Boyle traveled Europe with a private tutor studying, among many things, Galileo's work in Italy. Educated and refined, he lived the life of an aristocrat. As a young man he wrote on theology, morality, and ethics and continued to pursue these interests throughout his life. He soon became interested in natural philosophy and established a laboratory at his family estate in England. In a time when there existed no scientific profession, Boyle was fortunate that his own wealth allowed him the resources and leisure time to support his research. In his laboratory, Boyle developed an interest in experimental chemistry, and became determined to wrestle the discipline away from the scholastics and to make it a part of the emerging mechanical philosophy.

In 1655 Boyle moved to Oxford. Here he joined a group of young intellectuals interested in the new developments in natural philosophy. Boyle and other forward thinking men such as the mathematicians John Wilkins and John Wallis, along with the architect Christopher Wren, became interested in the Baconian program of empirical study. Initially meeting at various locations in London and Oxford, this invisible college eventually began meeting at Gresham College in London. The group was formalized in 1662 with a charter from King Charles II, and thus one of the earliest and most influential scientific societies was born — the Royal Society. When Boyle moved to London in 1668, his involvement with the Royal Society increased and the influence of this important institution spread.

Although the membership of the Royal Society was open to any gentleman interested in natural philosophy, its most important members were dedicated to a program of experimentation and speculation in the new sciences emerging in the seventeenth century. Robert Hooke, a gifted experimentalist who also served as Boyle's personal research assistant, was appointed curator of experiments by the society. In that capacity, Hooke designed and showcased mechanical experiments for the edification (and sometimes the entertainment) of the gathered Society Fellows. Together, Boyle and Hooke devised ingenious experiments with the air pump.

The air pump, recently invented by Otto von Guericke, was used to evacuate the air from a sealed vessel. Boyle immediately grasped the importance to science of such a device and, with the help of Hooke, began building his own air pumps. Hooke, whose patience and talent was required to make the temperamental devices work properly, performed a series of experiments with his employer, Boyle, to determine the nature of the vacuum. These experiments, published in Boyle's landmark treatise on experimental philosophy, *Experiments Physico–Mechanical, Touching the Spring of the Air and Its Effects*, showed (in addition to the fact that the vacuum really did exist), the effect of the vacuum on combustion, on respiration in animals, on magnetism, and on sound.

Boyle wrote about an experiment he conducted to determine the role of air in propagating sound. He described an experiment in which he hung a watch in a sealed glass globe. Hanging in the globe, the sound of the watch could clearly be heard by Boyle and others in the room. Boyle then proceeded to pump the air out of the globe, and described what happens:

> The pump after this being employed it seemed that from time to time the sound grew fainter and fainter, so that when the receiver was emptied; as much as it used to be for the foregoing experiments, neither we, nor some strangers that chanced to be then in the room could, by applying our ears to the very sides, hear any noise from within, though we could easily perceive that by moving of the hand which marked the second minutes, and by that of the balance, that the watch neither stood still, nor remarkably varied from its wonted motion. And to satisfy ourselves further that it was indeed the absence of the air about the watch that hindered us from hearing it, we let in the external air at the stop-cock, and then though we turned the key and stopped the valve, yet we could plainly hear the noise made by the balance, though we held our ears sometimes at two-foot distance from the outside of the receiver. And this experiment being reiterated in another place, succeeded after the like manner, which seems to prove that whether or not the air be the only, it is at least the principal medium of sounds.

This simple experiment (which Boyle is careful to note he replicated more than once) not only demonstrated that air is necessary for the propagation of sound, but also that a vacuum does exist, contrary to traditional opinions.

The role of air in the propagation of sound was only one of many experiments Boyle, with the assistance of Hooke, undertook. The two pioneering experimenters showed that air was necessary for combustion by removing air from a sealed vessel containing a burning candle. They proved that the same air that supported combustion was required for respiration by a series of experiments using small animals, such as birds, in the vessels. Boyle and Hooke found that as the air was pumped from the vessel the animal began to lose consciousness and would, of course, eventually die. However, if the seal was broken and air was allowed to reenter, the animal revived. All of these experiments indicated the "effects" of the "spring of air" as noted in the title of Boyle's work. But what exactly was this spring (or elasticity) of air and how can it be explained? Boyle wrote:

> Our air either consists of, or at least abounds with, parts of such a nature, that in case they be bent or compressed by the weight of the incumbent part of the atmosphere, or by any other body, they do endeavor, as much as in them lies, to free themselves from that pressure, by bearing against the contiguous bodies that keep them bent, and as soon as those bodies are removed or reduced to give them way, by presently unbending and stretching out themselves, either quite, or so far forth as the contiguous bodies that resist them will permit, and thereby expanding the whole parcel of air, these classical bodies compose.

Boyle depended once again on the theory of corpuscles (the parts of air) to explain his theory of the spring of air. Experiments with the barometer and the air pump, along with Boyle's theory of the elasticity of air, brought him to an important conclusion: the volume of a given quantity of air is inversely proportional to its pressure. This law, known as Boyle's law of gases, is one of the first laws of chemistry to be experimentally deduced.

Throughout his life, Boyle held fast to the mechanical philosophy, constructing his theories around the structural nature of corpuscles. In *On the Excellency and Grounds of the Corpuscular or Mechanical Philosophy*, Boyle argued incessantly that matter and motion (and matter in motion) are the basis from which all mechanical philosophy must be understood:

> I next observe that there cannot be fewer principles, than the two grand ones of our philosophy, matter and motion.... Nor can we conceive any principles more primary than matter and motion.... There cannot be any physical principles more simple than matter and motion; neither of them being resoluble into any other thing.

Boyle also argued for the universality of matter and motion, in the form of the smallest particles of matter — corpuscles:

> For the mechanical properties of matter are to be found, and the laws of motion take place, not only in the great masses and the middle-sized lumps, but in the smallest fragments of matter.... And therefore, to say that in natural bodies, whose bulk is manifest and their structure visible, the mechanical principles may be usefully admitted but are not to be extended to such portions of matter, whose parts and texture are invisible, is like allowing that the laws of mechanism may take place in a town-clock, and not in a pocket-watch ... for the mechanical principles are so universal, and applicable to so many purposes that they are rather fitted to take in, than to exclude, any other hypothesis founded on nature.

The mechanical philosophers looked to the characteristics of the corpuscles to explain physical properties of the bodies that they composed. A corpuscle of an acid was sharp and pointed, like a needle. Since these same acids reacted with alkalis to form innocuous salts, the alkaline corpuscles were perceived to be something like pin cushions, absorbing and softening the sting of the acids. Boyle described the properties of corpuscles of salt in a similar way:

> For the solidity, taste, etc., of salt may be fairly accounted for by the stiffness, sharpness, and other mechanical properties of the minute particles whereof salt consists.

The variety of shapes and characteristics of corpuscles led Boyle to an analogy between bricks as the building blocks of architects to corpuscles as the building blocks of nature:

> For if with the same bricks, differently put together and arranged, several bridges, vaults, houses, and other structures may be raised merely by various contrivances of parts of the same kind, what a great variety of ingredients may be produced by nature from the various coalition and contextures of corpuscles, that need not be supposed, like bricks, all of the same size and shape, but to have, both in the one and the other, as great a variety as could be wished for?

Regardless of the makeup of nature, the laws which govern natural phenomena are clear:

> By whatever principles natural things are constituted, it is by the mechanical principles that their phenomena must be clearly explained.

Although committed to the corpuscle theory, Boyle the experimentalist conceded that the building blocks of nature might be otherwise constituted. However, any principle used to explain the observed phenomena must be mechanical.

In *Certain Physiological Essays Written at Distant Times and on Several Occasions*, Boyle explained the properties of fluids by appealing to the characteristics of corpuscles:

A body then seems to be fluid, chiefly upon this account, that it consists of corpuscles that touching one another in some parts only of their surfaces, and being incontiguous in the rest, and separately agitated to and fro, can by reason of the numerous pores or spaces necessarily left between their incontiguous parts, easily glide along each other's superficies, and by reason of their motion diffuse themselves, till they meet with some hard or resisting body.

Contrast this description of the properties of fluids with those of solids:

We may conceive that the firmness of stability of a body consists principally in this, that the particles that compose it, besides that they are most commonly somewhat gross,* either do so rest or are so entangled between themselves, that there is among them a mutual cohesion whereby they are rendered unapt to flow or diffuse themselves every way, and consequently to be, without violence, bounded and figured by other surfaces than those which their connection makes themselves constitute.

The reasoning Boyle presented for fluidity and firmness is consistent with his other work — all physical phenomena may be explained by matter (in the form of corpuscles) and motion. Boyle presented his findings on the properties of fluids and solids through a series of experiments, each described in great detail.

Boyle published many books on the new chemistry, always advocating for the experimental method for obtaining results while employing the mechanical philosophy for explaining those results. Perhaps his most important work was entitled *The Skeptical Chymist*. In this book, Boyle published the results of many of his chemical experiments. In one of these experiments, Boyle reported that he separated potassium nitrate (or nitre as it was then known) into its components and then reconstituted the components back into potassium nitrate. Boyle argued that the results of this experiment — and many more like it — were proof of the corpuscle theory of matter.

In *The Skeptical Chymist*, Boyle launched an attack on the scholastics and would-be natural philosophers. He chastised them for their reliance on the word of ancient authorities over careful and repeatable experiments.

And as the obscurity of what some writers deliver makes it very difficult to be understood; so the unfaithfulness of too many others makes it unfit to be relied on. For though unwillingly, yet I must for the truths sake, and the readers, warn him not to be forward to believe Chymical Experiments when they are set down only by way of prescriptions, and not of relations; that is, unless he that delivers them mentions his doing it upon his own particular knowledge, or upon the relation of some credible person, avowing it upon his own experience. For I am troubled, I must complain, that even eminent writers, both physicians and philosophers, whom I can easily name, if it be required, have of late suffered themselves to be so far imposed upon, as to publish and build upon Chymical Experiments, which questionless they never tried; for if they had, they would, as well as I, have found them not to be true. And indeed it were to be wished, that now that those begin to quote Chymical Experiments that are not themselves acquainted with Chymical Operations, men would leave off that indefinite way of vouching the Chymists say this, or the Chymists affirm that, and would rather for each experiment they allege name the author or authors, upon whose credit they relate it; For, by this means they would secure themselves from the suspicion of falsehood (to which the other practice exposes them) and they would leave the reader to judge of what is fit for him to believe of what is delivered, whilst they employ not their own great names to countenance doubtful relations; and they will also do justice to the inventors or

*Large.

publishers of true experiments, as well as upon the obtruders of false ones. Whereas by that general Way of quoting the Chymists, the candid writer is defrauded of the particular praise, and the impostor escapes the personal disgrace that is due to him.

Boyle also sharply criticized the empty arguments of the scholastics over terminology such as elements and principles. Was all matter constituted of earth, water, air, and fire, as proclaimed by Aristotle so long ago? Or were there three principles of matter: salt, sulfur, and mercury, as argued by the Swiss physician and alchemist, Paracelsus? Boyle was above such rhetoric, defining an element in his own simple and straightforward way:

> Elements and Principles [may be considered] as terms equivalent: and to understand both by the one and the other, those primitive and simple Bodies of which the mixed ones are said to be composed, and into which they are ultimately resolved.

Boyle combined his interest in theology (he was a deeply devout man) with natural philosophy to explore the relationship between God and the natural world. Although a mechanical philosopher, Boyle believed that the properties of matter were imbued by God and rejected the notion of a creator who took no active part in nature. In *On the Excellency and Grounds of the Corpuscular or Mechanical Philosophy*, Boyle wrote:

> Thus, the universe was, once framed by God and the laws of motion, settled and everything was upheld by his perpetual concourse and general providence, the same philosophy teaches that the phenomena of the world are physically produced by the mechanical properties of the parts of matter, and that they operate upon one another according to mechanical laws.

This passage mirrors the beliefs of Descartes, who wrote in *The World*:

> From the first instant of their [fundamental particles] creation, he [God] causes some to start moving in one direction and others in another, some faster and others slower (or even, if you wish, not at all); and he causes them to continue moving, thereafter, in accordance with the ordinary laws of nature.

Indeed, although the goal of the mechanical philosophers was to remove the supernatural, occult, and animistic forces from nature, few went so far as to remove God from the equation. It was the creator who instilled in nature the fundamental laws governing matter and motion.

The Mechanical Philosophy

Descartes and Boyle were by no means the only natural philosophers of the period who embraced and extended the ideas of the mechanical philosophy. In fact, most of the important scientists of the seventeenth century (at least until the emergence of the work of Isaac Newton) were in one form or another committed to the mechanical philosophy. Christiaan Huygens, a Dutch mathematician, astronomer, and inventor, made contributions to the evolution of the mechanical philosophy, including correcting Descartes' theory on the conservation of momentum of colliding bodies and developing the wave theory of light. Huygens is best remembered today as the inventor of the pendulum clock, a fitting epitaph considering the widespread analogy used by mechanical philosophers comparing the universe to a clock.

The mechanical philosophy led some to conclusions that were unsettling to traditional anthropocentric thinking. An Italian anatomist, Giovanni Borelli, conceived of animals as machines, whose joints and other body parts operated as levers and pulleys. He expressed similar ideas concerning the human body, blurring the traditional distinction between man and beast. At about the same time, an Englishman named Edward Tyson showed with dissections that animals not outwardly similar were often more alike than first appeared. For instance, his dissections showed that a porpoise (historically thought to be a fish, for obvious reasons) was actually a mammal. He also studied chimpanzees, showing through dissection the similarities between chimps and humans. Borelli's theory of humans as machines and Tyson's comparative dissections of porpoises and chimps begin to blur the special place of man in the universe and encouraged the heretical thinking that the world was a machine indifferent to the existence of humans.

Proponents of the mechanical philosophy often took their ideas to such an extreme as to cause rifts among their fellow philosophers and with the establishment in the form of the universities or the church. The philosopher Thomas Hobbes, for example, conceived of humankind in purely mechanical terms, both body and soul. This sort of thinking led many to condemn the mechanical philosophers as materialists, emphasizing to an extreme the physical over the spiritual.

In spite of the prominence of Descartes and Boyle in the movement, there was really no single mechanical philosophy, but rather an array of ideas and theories gathered around a few central tenets. All mechanical philosophers agreed that nature could best be understood in terms of mechanics — matter in motion. Although some were atomists and embraced the existence of a vacuum, while others insisted with Descartes that a void could not exist, they nearly universally agreed that motion was best explained by direct contact between moving bodies. Another basic tenet of the mechanical philosophy was that the fundamental laws of nature were best understood through mathematical description. In spite of the fact that the mechanical philosophers were often accused of being materialists, and even atheists, most of them were devout and even rather orthodox Christians who maintained a central place for God in creating the very laws of nature they sought to discover.

Although much of what the mechanical philosophers had to offer was discarded by science after the seventeenth century, for a time the mechanical philosophy was the predominant theory of motion for all objects, living or not, in the universe. The most important legacy of the mechanical philosophy remains its commitment to explanatory structures built around reason, mathematics, and mechanics rather than spirits, animistic forces, and the occult.

PRIMARY SOURCES

Descartes, René. *Discourse on the Method of Rightly Conducting One's Reason and of Seeking Truth in the Sciences.* 1637. Project Gutenberg. http://www.gutenberg.org/files/59/59-h/59-h.htm.

_____. *The World. The World and Other Writings*, translated by Stephen Gaukroger. Cambridge: Cambridge University Press, 1998.

Boyle, Robert. *Certain Physiological Essays Written at Distant Times and on Several Occasions.*

_____. *Experiments Physico–Mechanical, Touching the Spring of the Air and Its Effects. The Scientific Revolution: A Brief History with Documents*, by Margaret C. Jacob. Boston: Bedford/St. Martin's, 2010.

_____. *On the Excellency and Grounds of the Corpuscular or Mechanical Philosophy. Scientific Revolutions: Primary Texts in the History of Science*, by Brian S. Baigrie. Upper Saddle River, NJ: Pearson Prentice Hall, 2004.

_____. *The Skeptical Chymist.* Project Gutenberg. http://www.gutenberg.org/files/22914/22914-h/22914-h.htm.

OTHER SOURCES

Aczel, Amir D. *Descartes' Secret Notebook: A True Tale of Mathematics, Mysticism, and the Quest to Understand the Universe*. New York: Broadway Books, 2005.

Anstey, Peter. *The Philosophy of Robert Boyle*. London: Routlege, 2000.

Ariew, Roger. *Descartes and the Last Scholastics*. Ithaca: Cornell University Press, 1999.

Collins, James Daniel. *Descartes' Philosophy of Nature*. Oxford: Blackwell, 1971.

Cottingham, John. *The Cambridge Companion to Descartes*. Cambridge and New York: Cambridge University Press, 1992.

_____. *Descartes*. New York: Oxford University Press, 1998.

Davis, J. Philip, and Rueben Hersch. *Descartes' Dream: The World According to Mathematics*. San Diego: Harcourt Brace Jovanovich, 1986.

Devlin, Keith J. *Goodbye Descartes: The End of Logic and the Search for a New Cosmology of the Mind*. New York: Wiley, 1997.

Gaukroger, Stephen. *Descartes: An Intellectual Biography*. Oxford: Clarendon Press; New York: Oxford University Press, 1995.

_____. *Descartes: Philosophy, Mathematics, and Physics*. Sussex: Harvester Press; Totowa, NJ: Barnes and Noble Press, 1980.

_____. *Descartes' System of Natural Philosophy*. Cambridge and New York: Cambridge University Press, 2002.

Hall, Marie Boas. *Robert Boyle and Seventeenth-Century Chemistry*. Cambridge: Cambridge University Press, 1958.

Hunter, Michael. *Boyle: Between God and Science*. New Haven: Yale University Press, 2009.

_____. *Robert Boyle (1627–91): Scrupulosity and Science*. Woodbridge, Suffolk, UK: The Boydell Press, 2000.

Principe, Lawrence M. *The Aspiring Adept: Robert Boyle and His Alchemical Quest*. Princeton: Princeton University Press, 1998.

Sargent, Rose-Mary. *The Diffident Naturalist: Robert Boyle and the Philosophy of Experiment*. Chicago: University of Chicago Press, 1995.

Schouls, Peter A. *Descartes and the Possibility of Science*. Ithaca: Cornell University Press, 2000.

Shapin, Steven, and Simon Schaffer. *Leviathan and the Air-Pump: Hobbes, Boyle and the Experimental Life*. Princeton: Princeton University Press, 1985.

Shea, William R. *The Magic of Numbers and Motion: The Scientific Career of René Descartes*. Canton, MA: Science History Publications, U.S.A., 1991.

Sorell, Tom. *Descartes*. Oxford; New York: Oxford University Press, 1987.

_____. *Descartes Reinvented*. Cambridge and New York: Cambridge University Press, 2005.

Voss, Stephen. *Essays on the Philosophy and Science of René Descartes*. New York: Oxford University Press, 1993.

Vrooman, Jack Rochford. *René Descartes; A Biography*. New York: Putnam, 1970.

Watson, Richard A. *Cogito Ergo Sum: The Life of René Descartes*. Boston: David R. Godine, 2002.

CHAPTER 6

Linnaeus, Buffon and Eighteenth Century Natural History

I wish it, therefore, to be acknowledged by all true botanists, if they ever expect any certainty in the science, that the *genera and species must be all natural.*— Carl Linnaeus, *The Genera of Plants* (1737)

The changes which the earth has undergone, during the last two or three thousand years, are inconsiderable, when compared with the great revolutions which must have taken place in those ages that immediately succeeded the creation.— Georges-Louis Leclerc, Comte de Buffon, *Natural History* (1749–1788)

The eighteenth century was an exhilarating time for Europeans. The Enlightenment burst upon the scene with optimistic ideas about humankind's abilities to know and understand nature. These beliefs manifested themselves in the physical sciences primarily through the influence of Isaac Newton (see Chapter 7). In the life and earth sciences, or natural history, several important leaders emerged in the eighteenth century. Two of them, the Swedish naturalist Carl Linnaeus and the French polymath Georges-Louis Leclerc, Comte de Buffon, were particularly influential. Linnaeus' greatest contributions came with his overhaul of the naming systems for both plants and animals, providing a stable, logical base from which botanists and zoologists could work. Buffon, on the other hand, left no particular legacy like Linnaeus. His influence was more wide-ranging and general, but his ideas on the antiquity of the earth and his empirical approach to natural history paved the way for future naturalists.

Taxonomy Before the Eighteenth Century

Before Linnaeus, taxonomy, or the system of classifying plants and animals, was a mishmash of ancient philosophy and colloquial naming schemes. Although Aristotle was the authority on taxonomy since antiquity, by the eighteenth century it could be said there were almost as many classification systems as there were scientists producing the systems. Common names for plants and animals, descriptive but decidedly superficial, were often anthropocentric. Today we can only imagine the properties of plants with names like stinking arrach or mare's fart. Plants with names such as maidenhair or shepherd's purse provided physical descriptions by way of human characteristics or products. Of course many of these anthropocentric names survive today. One simply needs to visit the local pharmacy to find St. John's Wort, or scan the shelves of the supermarket to find kidney beans. Descriptive as these names might be, naturalists realized that a scientific classification scheme required an approach that did not rely on simple outward appearances and characteristics. In addition, as naturalists and other scientists and intellectuals began to realize that the world may not have been created for the sole benefit of humans, the habit of naming things based on their benefit or relationship to humans began to change. The goal of scientists became finding a natural system that described and classified plants and animals in a way that addressed the essence of their structures.

One of these early innovators in classification was the English naturalist John Ray. Ray (1628–1705) wrote several books classifying thousands of species using a more natural system that focused on the overall similarities of plants. In his most important work, *The History of Plants*, Ray described over 18,000 different species. In the process, he used a definition for the term species that was the clearest yet to be written:

> After long and considerable investigation, no surer criterion for determining species has occurred to me than the distinguishing features that perpetuate themselves in propagation from seed. Thus, no matter what variations occur in the individuals or the species, if they spring from the seed of one and the same plant, they are accidental variations and not such as to distinguish a species.... Animals likewise that differ specifically preserve their distinct species permanently; one species never springs from the seed of another nor vice versa.

Ray was also important for his insistence that fossils were the remnants of previously living organisms. This theory about the true nature of fossils contradicted many long-standing interpretations which argued that fossils were just games of nature; rocks that, although resembling fish, or shells, or plants, and the like, were actually just created in that form. Ray's works had wide-spread influence on generations of naturalists, including the leading naturalist of the eighteenth century, Carl Linnaeus.

Carl Linnaeus

Born to a Lutheran pastor (and avid gardener) who passed on his love of nature to his son, Carl Linnaeus (1707–1778) spent his childhood exploring the countryside in his native Sweden. Young Linnaeus studied medicine at Uppsala University, although botany continued to be his true passion. While still a student at Uppsala, Linnaeus organized and directed an expedition to Lapland funded by a grant from the local scientific society. Linnaeus studied the flora and fauna of the region, as well as its geography and culture. Eventually granted a doctorate of medicine from Harderwijk University in Holland, Linnaeus juggled his medical career with his botanical and zoological interests throughout the rest of his life. In 1735, the same year he received his medical degree, Linnaeus published the first edition of his systematic study of the natural world, *The System of Nature*. While this first edition represented a brief foray into nature, Linnaeus continued to add to his work and republish expanded editions of *The System of Nature* until it eventually identified and named over 12,000 species of plants and animals.

Linnaeus began practicing medicine in Stockholm and was eventually appointed professor of medicine at Uppsala University. In addition to becoming the physician to the Swedish royal family, Linnaeus also made a significant contribution to Swedish intellectual life when he co-founded the Swedish Academy of Sciences. Throughout his life, Linnaeus mounted many expeditions to study the flora and fauna of various European regions. Perhaps more importantly, Linnaeus drew to Uppsala many devoted students who in turn embarked on voyages around Europe — indeed around the world — collecting specimens and spreading the techniques, theories, and philosophy of their teacher.

Carl Linnaeus died in 1778, leaving behind a legacy unmatched by any naturalist until Darwin. Commonly known as the "father of taxonomy," his long-lasting influence is the consistent use of a hierarchical system of naming based on physical characteristics shared by plants and by animals. Although the details — and in many cases even the underlying assumptions — have changed and evolved since the time of Linnaeus, his fundamental system established the architecture of taxonomy. Revered in his home country of Sweden, Linnaeus was also honored and admired throughout Europe. After his death, the British botanist James Edward Smith purchased Linnaeus' collections and founded the Linnean Society of London. The Linnean Society was, and continues to be, one of the world's leading natural history organizations, emphasizing the dissemination of taxonomic and conservation materials worldwide.

The Works of Linnaeus

Although Linnaeus introduced a wide variety of innovations in taxonomy, the two for which he is most famous are the sexual classification of plants and the binomial system of classification. Interestingly, Linnaeus did not invent the sexual system for naming plants; in fact, he readily credited others with the idea before him. In his *A Dissertation on the Sexes of Plants* (1760), Linnaeus went to great lengths to give credit to his predecessors, going so far as to point out that many ancient and medieval philosophers must have known that plants reproduce sexually. Referring to the "great chain of being," a philosophy dating to the ancient Greeks which connects all life in a hierarchical structure, Linnaeus maintained:

> To illustrate the generation of plants, we must take our first lights from the animal kingdom, and pursue the chain of nature until it leads us to vegetables.... Animals are, by the lower tribes of *Zoophyta*, brought so near to vegetables, that as I have before observed, we in some cases scarcely know how to distinguish the one from the other.

Linnaeus proceeded to use various analogies to animal sexual organs when describing the sex organs of plants. He backed up his conclusions with evidence from his own experiments on plant reproduction.

In writing about this taxonomical system in the tenth edition of *The System of Nature*, Linnaeus defined the reproductive organs of plants by noting, "Flowers that possess antlers are called *male*, those with stigmas *female*, and those which have both at a time *hermaphrodites*."

Linnaeus divided the plant kingdom into 24 classes based on sexual parts. In spite of his importance to modern taxonomy, Linnaeus often reverted to the anthropomorphic tendencies of his predecessors, using analogies to human sexuality when describing plants. For example, for class 8 (plants with 8 stamens clustered around 1 pistil), Linnaeus described "eight men in the same bridal suite with one woman." As one would expect, such allusions often offended eighteenth century sensibilities.

Just as describing plant reproduction sexually was morally repugnant to some, Linnaeus classification of man as part of the animal kingdom was roundly criticized and attacked. As Linnaeus contemplated humankind's place in the universe, he came to realize that the differences between humans and primates were minute. For this reason, Linnaeus placed humans alongside other creatures in the animal kingdom, placing them in the order *primates* (along with apes and bats) and the genus *homo* (along with orangutans). In a letter defending his work, Linnaeus lamented:

> It is not pleasing that I placed humans among the primates, but man knows himself. Let us get the words out of the way. It will be equal to me by whatever name they are treated. But I ask you and the whole world a generic difference between men and simians in accordance with the principles of Natural History. I certainly know none. If only someone would tell me one! If I called man an ape or vice versa I would bring together all the theologians against me. Perhaps I ought to scientifically....

Traditionally, humans held a special place in the universe, a separate creation. By having the temerity to face the criticism from all sides, both scientific and religious authorities, Linnaeus paved the way for others to consider further the place of humans in the universe.

Although Linnaeus' life-long goal was to discover a natural classification system, he acknowledged that even in this he comes up short: "No natural system of plants, though one or the other approaches it quite closely, has so far been constructed; nor do I contend that this system is really natural." However, Linnaeus concluded "Every genus is natural, thus created in the very beginning."

This last line reflects Linnaeus' goal of observing nature in order to uncover God's creation plan. He wrote:

> If we observe God's works, it becomes more than sufficiently evident to everybody, that each living being is propagated from an egg and that every egg produces an offspring closely resembling the parent. Hence no new species are produced nowadays.... As there are no new species (1); as like always gives birth to like (2); as one in each species was at the beginning of the progeny (3); it is necessary to attribute this progenitorial unity to some Omnipotent and Omniscient Being, namely *God*, whose work is called the *Creation*. This is confirmed by the mechanism, the laws, principles, constitutions and sensations in every living individual.

We can see from the previous paragraphs that Linnaeus was no evolutionist — he was sure of the fixity of the species, at least until late in his life when he began to think about species adapting to their environment. The next generation of naturalists, however, found much in Linnaeus' work that pointed them towards evolutionary theory, including ideas concerning competition and the struggle to survive, as well as thoughts on hybridization that hinted at the rise of new genera.

Linnaeus developed some very specific ideas concerning the role and scope of the naturalist. He believed that botany had a two-fold purpose—to organize and to name. In *The System of Nature*, he specified this role:

> The first step in wisdom is to know the things themselves; this notion consists in having a true idea of the objects; objects are distinguished and known by classifying them methodically and giving them appropriate names. Therefore, classification and name-giving will be the foundation of our science.

He later repeated this definition, specifying its application to botany: "The foundation of botany consists of the division of plants and systematic name-giving, generic and specific."

Linnaeus echoed this very practical goal for botanists in other works. In *The Species of Plants* (1753), he wrote:

> We have not resource, since we are not the governors of nature, nor can create plants according to our own conceptions, but to submit ourselves to the laws of nature, and learn by diligent study to read the characters inscribed on plants.... I acknowledge no authority but inspection alone in botany.

By refusing to acknowledge authorities, in his field, Linnaeus placed botany firmly in the spirit of the Enlightenment and followed his intellectual predecessors like Descartes and Harvey in shifting the responsibility of future scientists towards observation and induction.

Although well known for his use of the sexual system for classifying plants, Linnaeus' most important and influential contribution to natural history was the binomial naming system. Linnaeus introduced his binomial system of naming plants in *Botanical Philosophy* (1751), and in *The Species of Plants* published two years later he perfected his system. It would be another five years, however, with the 10th edition of *The System of Nature*, before Linnaeus applied the binomial system to the animal kingdom. Linnaeus readily acknowledged that the binomial system had been used by others before he adopted it, but the Swedish naturalist was the first to consistently and systematically apply the naming system to the natural world.

The binomial classification scheme, popularized by Linnaeus in his life-long quest for a natural system, identified plants and animals by genus and species. Although Linnaeus recognized that species fell short of his goal, he believed that the genera of plants did represent a natural description. In *The Genera of Plants* (1787), Linnaeus argued:

> The *genera* are as numerous as the common proximate attributes of the different species, as they are created in the beginning; this is confirmed by revelation, discovery, observation; hence the genera are all natural.... I do not deny that natural classes may not be given as well as natural genera. I do not deny that a natural method ought much to be preferred to ours, or those of other discoverers; but I laugh at all the natural methods hitherto cried up.... Let us nevertheless study plants, and in the mean time content ourselves with artificial and succedaneous classes.... These natural genera assumed two things are required to preserve them

pure; First, that the true species, and no others, be reduced to their proper genera. Secondly, that all the genera be circumscribed by true limits or boundaries, which we term *generic characters*.

Linnaeus "laughs" at the so-called natural methods of his predecessors.

Like many of the men who were instrumental in the scientific revolution, Linnaeus was not shy about meeting his critics head on. A young Linnaeus, not yet 30, published a work he called *A Critique of Botany* in 1737 in which he revealed the philosophy that guided his work throughout his career. In the dedication to the German born naturalist Johann Jacob Dillenius — then professor of botany at Oxford — Linnaeus sought the support of a more established scholar against the expected criticisms of his work:

> In publishing my observations concerning names I set your illustrious name at the head of my little book, that I may not be overwhelmed by the criticisms of malicious persons, who are more anxious to find some way of ridiculing the opinions of others than to make any original contribution to science themselves....

Impatient to have his work known to the world, and confident in his own methods and conclusions, Linnaeus declared that he "was unwilling ... to keep back the work merely to avoid the shafts of malevolence.... For I knew that wiser men ... do not fall under the spell of meretricious language."

Linnaeus commented upon the plethora of naming systems, and the fact that botanists were beginning to revolt against the chaos. He called the present state of taxonomy a "vast horde of names" and lamented that botanists agreed "anyone who should in future dare to introduce new names was stigmatized with a black mark."

After naming over a dozen botanists who had attempted to clean up the mess that was botanical nomenclature, Linnaeus concluded,

> and so also it is fated that botanists should impose wrong names, so long as the science remains an untilled field, so long as laws and rules have not been framed on which they [can] erect as on firm foundations the science of botany; and so the aforesaid botanists have, under pressure of necessity, corrected most wisely the faulty names given by their predecessors.

Later, Linnaeus compared his critics to "despots" who "busy themselves with gaining honor and authority for themselves from the disasters of others ... who, like pygmies taking their stand on the shoulders of giants, boast that they can see further."

After declaring that in his present work he would cite only "wiser authorities ... of superior learning ... [who] preferred the advancement of botany to every other consideration," Linnaeus took one last parting shot at his critics: "However, if I should wound other botanists of inferior rank, I ask their pardon, having set this down not out of malice but guided by my love of botany."

Even as a young man, Linnaeus had a clear vision of his life's work and of what botany should be. Commenting upon the present lack of clear and acceptable nomenclature in botany and the need for laws of naming, he wrote:

> How great a burden has been laid on the shoulders of botanists by disagreement in names, which is the first step toward barbarism.... Name changes have, however, hitherto been unavoidable among botanists, so long as no laws have been adopted by which names could be judged. Botanists live in a free state, and for them no eternal law can be prescribed unless it be adopted by the citizens both present and future, and indeed unless it be none other

than a law that can be shown by argument and example to be so faultless and indispensable that none better can be devised. Before botany can have such laws as I conceived and desire, the various citizens must advance their own arguments, and then let posterity decide on the best.

Clearly seeing himself as the one to fix the problems in botany, Linnaeus continued:

Before botanists can admit such laws it is necessary that someone among them should take upon himself to offer proposals to be examined by other botanists, so that if they are good they may be confirmed, if unsound they may be convicted of unsoundness and abandoned, while something better is put in their place. Now as hitherto no one has thought fit to undertake this self-denying task, I have determined to make this attempt.

The essence of Linnaeus life work was an effort to fix a broken system. Much like astronomy before Copernicus, botany — indeed all of natural history — was in disarray because of the ad hoc nature of the fixes that had been applied to taxonomy by centuries of naturalists. In language reminiscent of Copernicus' famous description of the "monstrous" situation in astronomy he sought to correct, Linnaeus wrote of his predecessors:

For some assuming the different parts of fructification as the principle of their system, and descending according to the laws of division from classes through orders even to species, broke and dilacerated the natural genera and did violence to nature by their hypothetical and arbitrary principles.

The results of these "hypothetical and arbitrary principles" led to the debacle that was classification in botany:

Hence arose many false genera; such controversy among authors; so many bad names, and such confusion! Such, indeed, was the state of things that as often as a new system-maker arose the whole botanic world was thrown into panic.

Linnaeus, like Copernicus before him, found a confusing, messy, and untenable system and dedicated his life to ending the confusion.

Buffon

Georges-Louis Leclerc, Comte de Buffon (1707–1788) was a French naturalist and mathematician whose influence on eighteenth century natural history extended at least as far as his Swedish contemporary and counterpart Linnaeus. As director of the Paris botanical gardens (the *Jardin des Plantes*), Buffon's influence on French science was unmatched in the eighteenth century. His monumental *Natural History*, published in 36 volumes and spanning a lifetime of work, attempted to catalog and categorize the entire natural world. Buffon's influence on the next generation of naturalists extended beyond the naming systems of Linnaeus into cosmology and natural philosophy.

Like many sons of prominent families, Buffon was sent to the university to study law. However, he found his interests lay more in science and mathematics and embarked on a career that would prove very fruitful. Buffon's first success came in mathematics, where he addressed a problem from a game the French called franc-carreau. The game itself was simple: a coin is dropped on a tiled floor with players betting whether the coin will come to rest completely within one tile or end up lying across two or more tiles. Buffon applied calculus and geometry to probability calculations to determine the likelihood of the various outcomes. Later, Buffon refined his techniques to include objects of various shapes, finally settling on a needle tossed onto a tile floor. Buffon's needle problem is considered a classic in probability theory.

Talented as he was in mathematics, Buffon is best remembered as an influential naturalist. Buffon's *Natural History* was one of the most widely read books of the eighteenth century. In spite of its 36 volume length, Buffon's style engaged the imagination of educated people in France, and indeed throughout Europe. The scope of *Natural History* was immense, with much of the work consisting of descriptions of animal species and mineralogical and geological observations. Conceptually, *Natural History* laid out a number of ideas that influenced naturalists for generations. One observation made by the French naturalist — that similar environments in different geographical locations contained different flora and fauna — is called today Buffon's Law. This and similar observations led to Buffon being called the "father of biogeography."

Buffon disputed the classification system of Linnaeus, and, indeed, any classification system that emphasized outward characteristics. In fact, Buffon doubted that any classification system could be valid since there really existed only individuals in nature, and these individuals were connected in "the Great Chain of Being" beginning at the top with humans and descending continuously to the simplest organisms. For Buffon, the concept of species centered upon the reproductive history of the organism:

> We should regard two animals as belonging to the same species if, by means of copulation, they can perpetuate themselves and preserve the likeness of the species; and we should regard them as belonging to different species if they are incapable of producing progeny by the same means. Thus the fox will be known to be a different species from the dog, if it proves to be the fact that from the mating of a male and a female of these two kinds of animals no offspring is born; and even if there should result a hybrid offspring, a sort of mule, this would suffice to prove that fox and dog are not the same species — inasmuch as this mule would be sterile.

One of Buffon's claims in *Natural History* engendered a dispute with Thomas Jefferson over the relative merits of the fauna of the Old and New Worlds. Buffon wrote:

> The immense territories of the New World contained not, upon its first discovery, a greater number of inhabitants than what are to be found in one half of Europe. This scarcity of the human species allowed the other animals to multiply prodigiously. They had fewer enemies and more space: Every circumstance was favorable to their increase; and each species, accordingly, consisted of a vast number of individuals. But the number of the species, when compared with those of the Old Continent, was not above one fourth, or one third. If we reckon that 200 species of quadrupeds exist in the whole known quarters of the globe, we shall find above 130 of them in the Old Continent, and less than 70 in the New; and, if we subtract the species common to both Continents, or those which, by their constitution, were able to endure the rigors of the North, and passed by land from the one Continent to the other, the New World cannot claim above 40 native species.

Buffon defended an interesting theory to which he attributes these differences in the number and vitality of species between the "Old Continent" and the "New."

In America, therefore, animated Nature is weaker, less active, and more circumscribed in the variety of her productions; for we perceive, from the enumeration of the American animals, that the numbers of species is not only fewer, but that, in general, all the animals are much smaller than those of the Old Continent. No American animal can be compared with the elephant, the rhinoceros, the hippopotamus, the dromedary, the camelopard, the buffalo, the lion, the tiger, &c. The tapir or *tapiierette* of Brazil, is the largest quadruped of South America. This animal, the elephant of the New World, exceeds not the size of a calf of six months old, or of a very small mule.

Of course these words drew the ire of many Americans, not the least of which was Thomas Jefferson, who, in addition to his well-known political career, also maintained a long-term interest in science, especially natural history. In fact, Jefferson's monograph *Notes on the State of Virginia* (1787), was considered one of the most important American books of the nineteenth century. In the book, Jefferson discussed the flora, fauna, geology, geography, anthropology, economy, industry, and even politics of his native state. Along the way, he also attempted to refute Buffon's claim that the New World had somehow produced "degenerate" animals. Buffon, Jefferson reminded his readers, had claimed that "nature is less active, less energetic on one side of the globe than she is on the other." Jefferson rebutted Buffon's theory with more than a hint of sarcasm:

> As if both sides were not warmed by the same genial sun; as if a soil of the same chemical composition, was less capable of elaboration into animal nutriment; as if the fruits and grains from that soil and sun, yielded a less rich chyle, gave less extension to the solids and fluids of the body, or produced sooner in the cartilages, membranes, and fibres, that rigidity which restrains all further extension, and terminates animal growth. The truth is, that a Pigmy and a Patagonian, a Mouse and a Mammoth, derive their dimensions from the same nutritive juices.

Jefferson produced tables comparing the relative size of species (both wild and domesticated) common to the Americas and Old World. Jefferson claimed his tables were not meant to "produce a conclusion in favour of the American species, but to justify a suspension of opinion until we are better informed, and a suspicion in the mean time that there is no uniform difference in favour of either." Jefferson's arguments must have been convincing, as Buffon eventually changed his mind and abandon his claim of the inferiority of the New World.

Jefferson closed his defense of the New World by addressing the accusations of the French intellectual, Abbé Raynal, that even white Europeans experienced some sort of degeneration after migrating to the Americas. In particular, Raynal pointed out, America had yet to produce a poet, a mathematician, an artist, or a scientist equal to those found throughout the history of Europe. Jefferson responded with the obvious — the United States' brief existence as a nation made the comparison unfair:

> When we shall have existed as a people as long as the Greeks did before they produced a Homer, the Romans a Virgil, the French a Racine and Voltaire, the English a Shakespeare and Milton, should this reproach be still true, we will enquire from what unfriendly causes it has proceeded, that the other countries of Europe and quarters of the earth shall not have inscribed any name in the roll of poets.

Even in the brief existence of America, Jefferson noted, a military genius (Washington) and a scientific genius (Franklin) have already appeared. Furthermore, John Rittenhouse, an American astronomer and master clock builder, must be considered in the first rank of genius "because he is self-taught." Finally, Jefferson argued that as a young country Americans were on the verge of producing such intellectual giants:

> As in philosophy and war, so in government, in oratory, in painting, in the plastic art, we might shew that America, though but a child of yesterday, has already given hopeful proofs of genius, as well of the nobler kinds, which arouse the best feelings.

Interestingly, Jefferson was not the only founding father who thought about the implications of building a new nation on the life of the mind. John Adams eloquently echoed Jefferson's sentiments that greatness in the arts and sciences would eventually spring from America. His argument was simply that at the present time the best American minds were occupied with more pressing needs. In a letter (1780) to his wife, Abigail, the future president wrote:

> I must study politicks and war that my sons may have liberty to study mathematicks and philosophy. My sons ought to study mathematicks and philosophy, geography, natural history, naval architecture, navigation, commerce and agriculture, in order to give their children a right to study painting, poetry, musick, architecture, statuary, tapestry and porcelaine.

Genius in the arts and sciences, both Jefferson and Adams strongly believed, would come in good time. Even if the new world was currently inferior in some way to the old — as Buffon claimed — the situation was temporary and soon changed.

One of the most important and long-lasting contributions made by Buffon to science was his questioning of the traditional explanation of the history of the earth. In a time when the creation and subsequent history of the world was almost undisputedly believed to be found in Genesis, Buffon wondered whether God might have used a natural mechanism for his creation.

> May we not conjecture, that a comet falling into the body of the sun might drive off some parts from its surface, and communicate to them a violent impulsive force, which they still retain? ... This effect was produced at the time when God is said by Moses to have separated the light from darkness ... on our supposition, there was a real physical separation; because the opaque bodies of the planets were detached from the luminous matter of which the sun is composed.

Buffon did not stop, however, with an unsubstantiated conjecture. He presented mathematical and observational evidence to back his conclusion. First, Buffon wondered, what was the probability that all of the planets orbited the sun in the same direction, if such things were left to chance?

> This notion concerning the cause of the centrifugal force of the planets will appear to be less exceptionable, after we have collected the analogies, and estimated the degrees of probability by which it may be supported. We shall first mention, that the motion of the planets have one common direction, namely, from west to east. By the doctrine of chances, it is easy to demonstrate, that this circumstance makes it as 64 to 1, that the planets could not all move in the same direction, if their centrifugal forces had not proceeded from the same cause.

Next, Buffon tackled the question of why all the planets should orbit in the same general plane.

> This probability will be greatly augmented, if we take in the similarity in the inclinations of the planes of their orbits, which exceed not 7½ degrees; for, by calculations it has been discovered, that it is 24 to 1 against any two planets being found, at the same time, in the most

distant parts of their orbits; and, consequently, 24⁵, or 7692624 to 1, that this effect could not be produced by accident; or, what amounts to the same, there is this great degree of probability, that the planets have been impressed with one common moving force, from which they have derived this singular position. But nothing could bestow this common centrifugal motion, excepting the force and direction of the bodies by which it was originally communicated.

Both of these arguments led Buffon to the conclusion that a comet impacting the sun created our solar system.

> We may, therefore, conclude, that all the planets have probably received their centrifugal motion by one single stroke. Having established this degree of probability, which almost amounts to a certainty, I next inquire what moving bodies could produce this effect; and I can find nothing but comets capable of communicating motion to such vast masses.

The next logical question, at least to Buffon, was how did sections of the sun become our modern planets? In particular, how long would it take such molten hot material to cool to a degree that would allow life? In order to answer this question, Buffon devised a series of experiments in which he heated iron balls (and later other materials) to extreme temperatures and then noted the time it took these balls to cool to ambient temperature. Using this data, he extrapolated to determine how long it should take a ball the size of the earth to cool sufficiently to allow the formation of life. His conclusion of almost 75,000 years was in direct conflict with the commonly held assumption that — based on the generations outlined in Genesis — the earth was only about 6000 years old. Although Buffon grossly underestimated the age of the earth, he became one of the first scientists to open the question of the world's history to experimentation. He also privately conjectured that the earth might indeed be much older than the 75,000 years he claimed in print. Buffon hinted at the antiquity of the earth, as well has humankind's relatively recent arrival on the scene:

> The changes which the earth has undergone, during the last two or three thousand years, are inconsiderable, when compared with the great revolutions which must have taken place in those ages that immediately succeeded the creation.... It appears, indeed, to be an incontrovertible fact, that the dry land which we now inhabit, and even the summits of the highest mountains, were formerly covered with the waters of the sea; for shells, and other marine bodies, are still found upon the very tops of mountains.... They must have been gradual and successive, as sea-bodies are sometimes found more than 1000 feet below the surface. Such a thickness of earth or of stone could not be accumulated in a short time. Although it should be supposed, that, at the deluge, all the shells were transported from the bottom of the ocean, and deposited upon the dry land; yet, beside the difficulty of establishing this supposition, it is clear, that, as shells are found incorporated in marble, and in the rocks of the highest mountains, we must likewise suppose, that all these marbles and rocks were formed at the same time, and at the very instant when the deluge took place; and that, before this grant revolution, there were neither mountains, nor marbles, nor rocks, nor clays, nor matter of any kind similar to what we are now acquainted with, as they all, with few exceptions, contain shells, and other productions of the ocean. Besides, at the time of the universal deluge, the earth must have acquired a considerable degree of solidity, by the action of gravity for more than sixteen centuries. During the short time the deluge lasted, it is, therefore, impossible, that the waters should have overturned and dissolved the whole surface of the earth, to the greatest depths that mankind have been able to penetrate.

First introduced in *Natural History*, Buffon's theories of the antiquity of the earth were fleshed out in his very popular (and controversial) *Epochs of Nature* (1788). In this work, Buffon related more details about his theory, dividing the history of the world into seven epochs. For instance, the first epoch began with the comet impacting the sun and ended with the earth cooling to the point of consolidation. The ensuing epochs found the formation of land masses and the atmosphere, the appearance and migration of plants and animals, and eventually — very recently in the earth's history — the appearance of humans.

Although today Buffon's theories of the formation and history of the earth seem far-fetched, it is important to realize that Buffon led a group of naturalists and scientists who first began questioning common assumptions about the earth's history. With Buffon's work, and later that of James Hutton, historical geology was born and scientists began to ascertain that the earth indeed had a long history — one in need of study beyond the chronological details of Genesis.

Although much of the work of Linnaeus and Buffon was supplanted in the nineteenth century, the naturalists of the eighteenth century paved the way for what was to come. Taxonomists continued to search — with varying degrees of success — for natural systems to categorize plants and animals. Earth scientists continued to investigate the history of the earth and its inhabitants. And scientists from all fields began to think about the implications of a universe that was almost unimaginably old.

PRIMARY SOURCES

Buffon, Georges-Louis Leclerc. *Natural History*, translated by William Smellie. http://faculty.njcu.edu/fm oran/buffonhome.htm.

Jefferson, Thomas. *Writings: Autobiography / Notes on the State of Virginia / Public and Private Papers / Addresses / Letters*, edited by Merrill D. Peterson. New York: Literary Classics, 1984.

Linnaeus, Carl. *Critica Botanica. The Autobiography of Science*, by Forest Ray Moulton and Justus J. Shifferes. New York: Doubleday, 1945.

_____. *A Dissertation on the Sexes of Plants*, translated by J.E. Smith. London, 1786.

_____. *The Genera of Plants. Scientific Revolutions: Primary Texts in the History of Science*, by Brian S. Baigrie. Upper Saddle River, NJ: Pearson Prentice Hall, 2004.

_____. *The System of Nature*. Translated by M.S.J. Engel-Ledeboer and H. Engel. Nieuwkoop: B. de Graaf, 1964.

OTHER SOURCES

Blunt, Wilfrid, and William T. Stearn. *The Compleat Naturalist: A Life of Linnaeus*. New York: Viking Press, 1971.

Dickinson, Alice. *Carl Linnaeus; Pioneer of Modern Botany*. New York: F. Watts, 1967.

Fara, Patricia. *Sex, Botany & Empire: The Story of Carl Linnaeus and Joseph Banks*. New York: Columbia University Press, 2003.

Farber, Paul L. *Finding Order in Nature: The Naturalist Tradition from Linnaeus to E.O. Wilson*. Baltimore: Johns Hopkins University Press, 2000.

Frängsmyr, Tore, and Sten Lindroth. *Linnaeus, the Man and His Work*. Canton, MA: Science History Publications, U.S.A., 1994.

Goerke, Heinz. *Linnaeus*. New York: Scribner, 1973.

Keith, Thomas. *Man and the Natural World: Changing Attitudes in England, 1500–1800*. Oxford: Oxford University Press, 1983.

Koerner, Lisbet. *Linnaeus: Nature and Nation*. Cambridge: Harvard University Press, 1999.

Roger, Jacques. *Buffon: A Life in Natural History*, translated by Sarah Lucille Bonnefoi. Ithaca: Cornell University Press, 1997.

Stoutenburg, Adrien, and Laura Nelson Baker. *Beloved Botanist: The Story of Carl Linnaeus*. New York: Scribner, 1961.

Weinstock, John M. Lanham. *Contemporary Perspectives on Linnaeus*. Lanham, MD: University Press of America, 1985.

CHAPTER 7

Newton and the Pinnacle
of the Scientific Revolution

I do not know what I may appear to the world, but to myself I seem to have been only like a boy playing on the sea-shore, and diverting myself in now and then finding a smoother pebble or a prettier shell than ordinary, whilst the great ocean of truth lay all undiscovered before me.—Isaac Newton to Robert Hooke (1676)

When the eighteenth-century British poet Alexander Pope penned the lines,

> Nature and Nature's laws lay hid in night:
> God said, Let Newton be!
> And all was light,

he was affirming the sentiments of most of the world. Isaac Newton had unveiled the laws of nature hidden from humankind for so long. Newton's work — and legend — continues today to capture the attention of scientists, historians, theologians, and almost anyone interested in our quest for knowledge of the world and of ourselves.

Newton's Life

Isaac Newton was born on a small farm in the English countryside in 1642. His father, a relatively prosperous farmer, died before Isaac was born. When his mother remarried and moved away, young Isaac was left with his grandmother on the family farm. Dealing with what by all accounts was an unhappy childhood, Isaac was at best a disinterested student. He did, however, display a remarkable talent for building toys and mechanical devices. An uncle saw enough potential in the young man to encourage his mother to allow Isaac to attend university. His mother capitulated, probably because her son showed no interest or aptitude for running the family farm.

Newton entered Trinity College, Cambridge, at the age of 18. Although today this is the usual age for a young person to begin college, at the time Newton was several years older than the typical entering student. It appears also that Newton was required to work as a servant for other students in order to help pay his way at Cambridge. The course of study at Cambridge was dominated by the works of Aristotle; however, Newton was able to also study the mechanical philosophy of Descartes and Boyle, as well as the astronomy of Copernicus and Kepler.

Newton's introduction to higher mathematics, as told by one of his early biographers, sheds important light on to his abilities and enthusiasm for learning. It seems that Newton acquired a book on astrology at a fair in Cambridge. Attempting to understand the mathematical astronomy in the book, he bought a book on trigonometry. Realizing his geometry knowledge was hindering his progress, he obtained a copy of Euclid's *Elements*, the primary textbook for geometry for the past two thousand years. Newton read and re-read Euclid until he had mastered all of the propositions and their proofs, then back-tracked through his trigonometry text and eventually achieved his goal of understanding the astrology work. Newton, however, did not stop here. His mathematical curiosity piqued, he proceeded to study the most advanced mathematics of the day, including the analytic geometry recently invented by Descartes and various algebraic works. In a very short period, Newton had taught himself the most advanced mathematics of the time.

In the spring of 1665, Newton received his bachelor's degree from Cambridge. Still having not distinguished himself as a scholar, his future was uncertain. And what was to come next certainly did not — at least on the surface — bode well for his future. Simmering in the crowded slums of London, the Plague, or Black Death, spread quickly throughout the city and into other parts of England. By the summer of 1665, the spread of the deadly disease was such that everyone who had the means began fleeing the crowded cities for country estates. Cambridge University was closed and students sent home in the hopes of avoiding catastrophe. Young Isaac Newton, recently awarded a degree from Cambridge, was back on the family farm. Once again it seems that his mother tried to make a gentleman farmer out of her son, but once again Newton's disinterest blocked all attempts. Instead, Newton spent the next year and a half in his room thinking and working. During this time, which has been called Newton's *annus mirabilis*, or miracle year, the young scholar developed many of the ideas and theories that revolutionized the world of science.

Later in life, Newton recorded in his notebook his recollection of that time:

In the beginning of the year 1665 I found the method of approximating series and the rule for reducing the dignity of any binomial into such a series. In the same year, in May, I found the method of tangents of Gregory and Sulzius, and in November had the direct method of fluxions and in the next year in January had the theory of colours, and in May following I had entrance into the inverse method of fluxions. And in the same year I began to think of gravity extending to the orb of the moon, and having found out how to estimate the force with which a globe revolving within a sphere presses the surface of a sphere, from Kepler's rule of the periodical times of the planets being in a sesquialterate proportion of their distances from the centres of their orbs I deduced that the forces which keep the planets in their orbs must [be] reciprocally as the squares of their distances from the centres about which they revolve: and thereby compared the force requisite to keep the moon in her orb with the force of gravity at the surface of the earth, and found them answer pretty nearly. All this was in the two years of 1665 and 1666, for in those days I was in the prime of my age for invention, and minded mathematics and philosophy more than at any time since.

When Newton wrote "in those days I was in the prime of my age for invention," he was guilty of some understatement. He found "the rule for reducing the dignity of any binomial into such a series," which is the binomial series, a fundamental breakthrough in mathematics; "the direct method of fluxions," called differential calculus today, and then later "the inverse method of fluxions," which is the other branch of calculus (integral calculus); "the theory of colours," which he later published in the *Opticks*; and while thinking "of gravity extending to the orb of the moon ... deduced that the forces which keep the planets in their orbs must [be] reciprocally as the squares of their distances from the centres about which they revolve"— the universal law of gravitation. Not bad for a year and a half of sitting in one's room thinking. Newton spent much of the coming years developing and perfecting his ideas in these areas and others.

Upon returning to Cambridge in 1667, Newton's career and reputation began to rise at a rapid pace. A master's degree and several fellowships preceded his appointment to the Lucasian Chair of Mathematics — a far cry from the young man who served his fellow students as an undergraduate. This position, previously held by Newton's mentor, Isaac Barrow, provided Newton with a secure and respectable position from which to work. It did not require a great number of duties, leaving Newton with the time and freedom he needed to pursue his scientific interests. Ironically, Newton almost missed the opportunity to acquire this important position. At the time, all professorships at Cambridge required the holder to take holy orders and become ordained in the Anglican Church. Although a devout man, Newton held (secretly) several unorthodox views* that would have kept him from taking such an oath. He appealed to King Charles II and was excused from taking the orders.

Newton spent the next several decades perfecting his theories and publishing numerous groundbreaking and revolutionary works, including *Mathematical Principles of Natural Philosophy* (1687) and *Opticks* (1704). During these years Newton, often considered a recluse due to his single-minded concentration on his work, also emerged as a leader both in the scientific community and in the political realm. Newton served for many years as a member of Parliament, representing Cambridge University. Then, in 1696, while recovering from a possible mental breakdown due to overwork, Newton was offered and accepted the position of warden of the English mint. In this position, Newton oversaw a recoinage effort aimed at reducing the rampant counterfeiting of English coins. Later appointed master of the mint, the income from his position made Newton a wealthy man in his later life.

Not neglecting his previous interests in science, Newton became president of the Royal Society of London in 1703, a position which he held until his death in 1727. For his scientific accomplishments as well as his work in the mint, Newton was knighted by Queen Anne in 1705, becoming Sir Isaac Newton. When Sir Isaac died in 1727, he was laid to rest at Westminster Abbey, an honor accorded only to monarchs and the most illustrious of their British subjects.

---

*Newton was an Arian. One of the views held by this Christian sect is that Jesus was a divine, but created, being. This rejection of the orthodox belief in the holy trinity was ironic, since Newton was appointed to a professorship in Trinity College, Cambridge.

The *Principia*

Generally regarded as the most important scientific book ever written, *Philosophiæ Naturalis Principia Mathematica* (*Mathematical Principles of Natural Philosophy*) — usually referred to as simply the *Principia* — might never have been written if not for the tenacity of one of Newton's fellow natural philosophers.

By 1684, Edmond Halley (1656–1742) had already established himself as an important young astronomer. Although it would yet be another decade before Halley predicted the return of the comet of 1682 — and its continual appearance every 76 years thereafter — Halley's work in cataloging the stars of the southern hemisphere resulted in accolades from the British astronomical establishment. In 1684, Halley turned his attention to the question of whether Kepler's law of elliptical orbits was implied by an inverse square law. In other words, if the force acting on the planets was inversely proportional to the square of their distances, would elliptical orbits necessarily follow? Halley and others worked on this problem to no avail, until Halley decided to approach a talented but still little known mathematician at Cambridge with the problem.

Halley was surprised to hear Newton's answer: yes, an inverse square law does imply elliptical orbits — in fact, Newton claimed, he had proven this very theorem some time before. Although Newton could not find the proof for Halley at the time, he eventually produced this important result. Over the coming days and months, Halley was astounded to learn of other results that Newton had found but not yet published. It was only on Halley's consistent urging that Newton began to compile the work that became the *Principia*. In fact, Halley not only encouraged Newton in his writing, he served as proofreader and copyeditor and even paid to publish the *Principia* with his own funds. Although the *Principia* sprang from the mind of Newton, it might never have seen the light of day was it not for Halley.

When the *Principia* finally appeared in 1687, it was hailed as a groundbreaking scientific work. The title itself clearly pointed to Newton's great accomplishment: for the first time, someone had produced a complete work on the *mathematical principles* of *natural philosophy* (or physics). In his preface to the *Principia*, Newton clearly stated his desire to solidify the place of geometry in natural philosophy:

> Since the ancients (as we are told by *Pappas*), made great account of the science of mechanics in the investigation of natural things; and the moderns, lying aside substantial forms and occult qualities, have endeavored to subject the phenomena of nature to the laws of mathematics, I have in this treatise cultivated mathematics so far as it regards philosophy. The ancients considered mechanics in a twofold respect; as rational, which proceeds accurately by demonstration; and practical. To practical mechanics all the manual arts belong, from which mechanics took its name. But as artificers do not work with perfect accuracy, it comes to pass that mechanics is so distinguished from geometry, that what is perfectly accurate is called geometrical; what is less so, is called mechanical. But the errors are not in the art, but in the artificers. He that works with less accuracy is an imperfect mechanic; and if any could work with perfect accuracy, he would be the most perfect mechanic of all; for the description of right lines and circles, upon which geometry is founded, belongs to mechanics. Geometry does not teach us to draw these lines, but requires them to be drawn; for it requires that the learner should first be taught to describe these accurately, before he enters upon geometry; then it shows how by these operations problems may be solved. To describe right lines and

circles are problems, but not geometrical problems. The solution of these problems is required from mechanics; and by geometry the use of them, when so solved, is shown; and it is the glory of geometry that from those few principles, brought from without, it is able to produce so many things. Therefore geometry is founded in mechanical practice, and is nothing but that part of universal mechanics which accurately proposes and demonstrates the art of measuring. But since the manual arts are chiefly conversant in the moving of bodies, it comes to pass that geometry is commonly referred to their magnitudes, and mechanics to their motion.

Newton defined the science of rational mechanics, and laid out for the reader the subjects he covered in his book.

In this sense rational mechanics will be the science of motions resulting from any forces whatsoever, and of the forces required to produce any motions, accurately proposed and demonstrated. This part of mechanics was cultivated by the ancients in the five powers which relate to manual arts, who considered gravity (it not being a manual power, no otherwise than as it moved weights by those powers). Our design not respecting arts, but philosophy, and our subject not manual but natural powers, we consider chiefly those things which relate to gravity, levity, elastic force, the resistance of fluids, and the like forces, whether attractive or impulsive; and therefore we offer this work as the mathematical principles of philosophy; for all the difficulty of philosophy seems to consist in this — from the phenomena of motions to investigate the forces of nature, and then from these forces to demonstrate the other phenomena; and to this end the general propositions in the first and second book are directed. In the third book we give an example of this in the explication of the System of the World; for by the propositions mathematically demonstrated in the former books, we in the third derive from the celestial phenomena the forces of gravity with which bodies tend to the sun and the several planets. Then from these forces, by other propositions which are also mathematical, we deduce the motions of the planets, the comets, the moon, and the sea. I wish we could derive the rest of the phenomena of nature by the same kind of reasoning from mechanical principles; for I am induced by many reasons to suspect that they may all depend upon certain forces by which the particles of bodies, by some causes hitherto unknown, are either mutually impelled towards each other, and cohere in regular figures, or are repelled and recede from each other; which forces being unknown, philosophers have hitherto attempted the search of nature in vain; but I hope the principles here laid down will afford some light either to this or some truer method of philosophy.

So, we can see from Newton's preface, the *Principia* is a book of mathematics. But not simply the theoretical and esoteric mathematics of the Greek geometry, but applied mathematics.

Although it is well beyond the scope of this book, it is important to the modern reader to understand the mathematical nature of Newton's masterpiece. The following excerpt is typical of the material found in the *Principia*. For the non-mathematically inclined, it is not critical to read the passage carefully — simply note the geometric nature typical of the *Principia*.

Of the motion of bodies in eccentric conic sections.

PROPOSITION XI. PROBLEM VI.

If a body revolves in an ellipsis; it is required to find the law of the centripetal force tending to the focus of the ellipsis.

Let S be the focus of the ellipsis. Draw SP cutting the diameter DK of the ellipsis in E, and the ordinate Q*v* in *x*; and complete the parallelogram Q*x*PR. It is evident that EP is equal to the greater semi-axis AC: for drawing HI from the other focus H of the ellipsis parallel to EC, because CS, CH are equal, ES, EI will be also equal; so that EP is the half sum of PS, PI, that is (because of the parallels HI, PR, and the equal angles IPR, HPZ), of PS, PH, which taken together, are equal to the whole axis 2AC. Draw QT perpendicular to SP, and putting L for the principal latus rectum of the ellipsis (or for 2BC²/AC), we shall have L · QR to L · Pv as QR to Pv, that is, as PE or AC to PC; and L · Pv to GvP as L to Gv; and GvP to Qv² as PC² to CD²; and by (Corol. 2, Lem. VII) the points Q and P coinciding, Qv² is to Qx² in the ratio of equality; and Qx² or Qv² is to QT² as EP² to PF², that is, as CA² to PF², or (by Lem. XII) as CD² to CB². And compounding all those ratios together, we shall have L · QR to QT² as AC · L · PC² · CD², or 2CB² · PC² · CD² to PC · Gv · CD² · CB², or as, 2PC to Gv. But the points Q and P coinciding, 2PC and Gr are equal. And therefore the quantities L · QR and QT², proportional to these, will be also equal. Let those equals be drawn into SP²/QR, and L · SP² will become equal to SP² · QT² / QR. And therefore (by Corol. 1 and 5, Prop. VI) the centripetal force is reciprocally as L · SP², that is, reciprocally in the duplicate ratio of the distance SP. Q.E.D.

The *Principia* contains many of the most famous theories in physics, but the first thing most people envision when they hear the name Isaac Newton is an apple falling. Although it is certainly just a caricature of a historical event, the popular story of an apple falling on Newton's head and his sudden inspiration concerning gravity is probably rooted in some fact. On the farm in those fateful years while the plague ravaged England, young Newton began thinking about the forces that caused an apple to fall from a tree. What if, he mused, the same forces that caused the apple to accelerate as it plummeted to the ground caused the moon be drawn to the earth?—or the planets to be drawn to the sun? This insight led Newton on his quest to discover a mathematical law that applied to both apples falling from trees and the moon falling towards the earth.

Newton's universal law of gravitation, as it is known today, was never stated in its modern form by Newton. In the *Principia,* Newton revealed his discovery in his own terminology:

> If spheres be however dissimilar (as to density of matter and attractive force) in the same ration onwards from the centre to the circumference; but everywhere similar, at every given distance from the centre, on all sides round about; and the attractive force of every point decreases as the square of the distance of the body attracted: I say, that the whole force with which one of these spheres attracts the other will be inversely proportional to the square of the distances of the centres.

Essentially, Newton's law of universal gravitation maintains that every object in the universe attracts every other object with a force that is proportional to the masses of the objects and inversely proportional to the square of the distance between the objects. In modern mathematical notation, it says:

$$F = G \frac{m_1 m_2}{d^2}$$

The value of G, the universal constant, was not determined until over a century after Newton conceived of the law of gravitation. Universal is a key word in Newton's discovery: this is a law that applies to objects on earth as well as objects throughout the universe, a gigantic leap from the ancient concept that declared there existed a fundamental difference between the heavens and the earth and therefore the same laws did not apply in both locations.

Newton's publication of his law of gravity in the *Principia* opened a dispute with another natural philosopher, Robert Hooke. Hooke, the long-time curator of experiments for the Royal Society, had communicated to Newton in a letter his theory that the force of attraction between bodies followed an inverse square law. Hooke's theory was speculation until Newton supplied the mathematical proof. Newton and Hooke had previously been involved in a public dispute over their competing theories of light, so Newton may have been predisposed to dislike Hooke. Whatever the case, Hooke was not given credit by Newton for the idea, and the dispute became heated. The relationship between Hooke and Newton remained strained throughout most of their respective lifetimes. In a letter to Hooke, Newton did seem to offer a conciliatory note when he wrote, "If I have been able to see further, it was only because I stood on the shoulders of giants." On the other hand, after Hooke's death, a large portrait was lost during a move of the Royal Society coordinated by its new president, Isaac Newton. Some speculate the disappearance of his rival's portrait was not an accident.

Next to his law of gravitation, Newton is perhaps best remembered for the three laws of motion that form the foundation of classical physics and are dutifully memorized by every schoolchild. Once again, the modern formulation of these laws has evolved to the point that Newton's original statements may not seem familiar. Newton stated his first law of motion as:

> Every body perseveres in its state of rest, or of uniform motion in a right line, unless it is compelled to change that state by forces impressed thereon.

In more familiar terms, this law states that an object at rest will tend to stay at rest unless acted upon by an outside force while an object in motion will stay in motion in the same direction unless acted upon by an outside force. Today, this is commonly called the law of inertia.

Newton's second law of motion, as stated in the *Principia*, stated:

> The alteration of motion is ever proportional to the motive force impressed; and is made in the direction of the right line in which that force is impressed.

This law is usually stated today as one of the most famous formulas in science $F = ma$, where F is the force on a body of mass m being accelerated by a value a. Finally, the third law of motion appears in the *Principia* as:

> To every action there is always opposed an equal reaction; or the mutual actions of two bodies upon each other are always equal, and directed to contrary parts.

Stated in modern terms, this becomes the famous law, for every action there is an equal and opposite reaction.

Newton assigned credit for the first of these two laws to his predecessor, Galileo, reminding the reader of a famous result obtained by the Italian physicist:

> Hitherto I have laid down such principles as have been received by mathematicians, and are confirmed by abundance of experiment. By the first two Laws and the first two Corollaries,

Galileo discovered that the descent of bodies observed the duplicate ratio of the time, and that the motion of projectiles was in the curve of a parabola; experience agreeing with both, unless so far as these motions are a little retarded by the resistance of the air.

Although the law of gravitation and the laws of motion are the most famous results found in the *Principia*, they are by no means the only important discoveries made by Newton and communicated in his monumental work. For instance, with his universal law of gravitation, Newton provided a theoretical basis for Kepler's three laws of planetary motion. Kepler himself had derived his laws empirically — primarily from the data collected by his mentor, Tycho Brahe. Newton proved, for instance, that an inverse square law like his law of gravitation implied an elliptical orbit for the planets with the sun at one of the focal points. He established similar proofs for Kepler's other two laws.

Newton also showed that Kepler's three laws were incompatible with Descartes' theory of vortices. This was a deciding event in the ongoing competition between Newtonian gravitational theory and the mechanical theories of Descartes. Yet, the mechanical philosophers were quick to point out that without a mechanism to explain gravity, Newton was regressing back to the ancient reliance on the occult and supernatural. Newton's response was to admit that he had no physical explanation for gravity, and that any such explanation would simply be hypothesis, which has no place in experimental philosophy:

> But hitherto I have not been able to discover the cause of those properties of gravity from phenomena, and I frame no hypothesis; for whatever is not deduced from the phenomena is to be called an hypothesis; and hypotheses, whether metaphysical or physical, whether of occult qualities or mechanical, have no place in experimental philosophy.... And to us it is enough that gravity does really exist, and act according to the laws which we have explained, and abundantly serves to account for all the motions, of the celestial bodies, and of our sea.

We see from the last sentence of the above quote another accomplishment published in the *Principia*. Newton was able to use his theory of gravity to account for the tides of the earth's oceans. Galileo had incorrectly postulated that the tides were caused by the movement of the earth itself, something akin to water sloshing around in a moving bucket. Interestingly, ancient cultures had associated the tides with the occult effect of the distant planets, an explanation not to dissimilar to Newton's explanation that it is the moon's gravity that pulls on the oceans and causes the tides.

In the third and final section of the *Principia*, Newton began by explaining to the reader his intentions for the first two books, along with a guide to the reader on how to approach the *Principia* based on each reader's abilities and interests in the subjects:

> In the preceding Books I have laid down the principles of philosophy, principles not philosophical, but mathematical: such, to wit, as we may build our reasonings upon in philosophical inquiries. These principles are the laws and conditions of certain motions, and powers or forces, which chiefly have respect to philosophy; but, lest they should have appeared of themselves dry and barren, I have illustrated them here and there with some philosophical scholiums, giving an account of such things as are of more general nature, and which philosophy seems chiefly to be founded on; such as the density and the resistance of bodies, spaces void of all bodies, and the motion of light and sounds. It remains that, from the same principles, I now demonstrate the frame of the System of the World. Upon this subject I had, indeed, composed the third Book in a popular method, that it might be read by many; but afterward, considering that such as had not sufficiently entered into the principles could not

easily discern the strength of the consequences, nor lay aside the prejudices to which they had been many years accustomed, therefore, to prevent the disputes which might be raised upon such accounts, I chose to reduce the substance of this Book into the form of Propositions (in the mathematical way), which should be read by those only who had first made themselves masters of the principles established in the preceding Books: not that I would advise any one to the previous study of every Proposition of those Books; for they abound with such as might cost too much time, even to readers of good mathematical learning. It is enough if one carefully read the Definitions, the Laws of Motion, and the first three Sections of the first Book. He may then pass on to this Book, and consult such of the remaining Propositions of the first two Books, as the references in this, and his occasions, shall require.

Newton proceeded to lay out his "system of the world," exploring the consequences of universal gravitation on the movements of the heavenly bodies.

One of the most interesting — and important — sections of the *Principia* did not even appear in the first edition. In later editions, Newton added and subsequently modified his "Rules of Reasoning in Philosophy," in which he laid out his own guide to scientific discovery. This guide had a long-lasting influence on generations of scientists searching for a framework around which to build their own science. The rules are straightforward and easy to understand, but revolutionary in their approach to natural philosophy, or science.

RULE I.

We are to admit no more causes of natural things than such as are both true and sufficient to explain their appearances.

Each rule is complete and simply stated. However, for each rule Newton expounded on its meaning, giving examples and adding clarity to the rule:

To this purpose the philosophers say that Nature does nothing in vain, and more is in vain when less will serve; for Nature is pleased with simplicity, and affects not the pomp of superfluous causes.

Newton proceeded along a similar line for each of the next three rules.

RULE II.

Therefore to the same natural effects we must, as far as possible, assign the same causes.

As to respiration in a man and in a beast; the descent of stones in *Europe* and in *America*; the light of our culinary fire and of the sun; the reflection of light in the earth, and in the planets.

RULE III.

The qualities of bodies, which admit neither intension nor remission of degrees, and which are found to belong to all bodies within the reach of our experiments, are to be esteemed the universal qualities of all bodies whatsoever.

For since the qualities of bodies are only known to us by experiments, we are to hold for universal all such as universally agree with experiments; and such as are not liable to diminution can never be quite taken away. We are certainly not to relinquish the evidence of experiments for the sake of dreams and vain fictions of our own devising; nor are we to recede from the analogy of Nature, which is want to be simple, and always consonant to itself. We no other way know the extension of bodies than by our senses, nor do these reach it in all bodies; but because we perceive extension in all that are sensible, therefore we ascribe it universally to all others also. That abundance of bodies are hard, we learn by experience; and because the hardness of the whole arises from the hardness of the parts, we therefore justly infer the hardness of the undivided particles not only of the bodies we feel but of all others. That all bodies are impenetrable, we gather not from reason, but from sensation. The bodies which we handle we find impenetrable, and thence conclude impenetrability to be an universal property of all bodies whatsoever. That all bodies are moveable, and endowed with certain powers (which we call the *vires inertiæ)* of persevering in their motion, or in their rest we only infer from the like properties observed in the bodies which we have seen. The extension, hardness, impenetrability, mobility, and *vis inertiæ* of the whole, result from the extension hardness, impenetrability, mobility, and *vires inertiæ* of the parts; and thence we conclude the least particles of all bodies to be also all extended, and hard and impenetrable, and moveable, and endowed with their proper *vires inertiæ.* And this is the foundation of all philosophy. Moreover, that the divided but contiguous particles of bodies may be separated from one another, is matter of observation; and, in the particles that remain undivided, our minds are able to distinguish yet lesser parts, as is mathematically demonstrated. But whether the parts so distinguished, and not yet divided, may, by the powers of Nature, be actually divided and separated from one another, we cannot certainly determine. Yet, had we the proof of but one experiment that any undivided particle, in breaking a hard and solid body, offered a division, we might by virtue of this rule conclude that the undivided as well as the divided particles may be divided and actually separated to infinity.

Lastly, if it universally appears, by experiments and astronomical observations, that all bodies about the earth gravitate towards the earth, and that in proportion to the quantity of matter which they severally contain, that the moon likewise, according to the quantity of its matter, gravitates towards the earth; that, on the other hand, our sea gravitates towards the moon; and all the planets mutually one towards another; and the comets in like manner towards the sun; we must, in consequence of this rule, universally allow that all bodies whatsoever are endowed with a principle of mutual gravitation. For the argument from the appearances concludes with more force for the universal gravitation of all bodies that for their impenetrability; of which, among those in the celestial regions, we have no experiments, nor any manner of observation. Not that I affirm gravity to be essential to bodies: by their *vis insita* I mean nothing but their *vis inertiæ.* This is immutable. Their gravity is diminished as they recede from the earth.

RULE IV.

In experimental philosophy we are to look upon propositions collected by general induction from phenomena as accurately or very nearly true, notwithstanding any contrary hypotheses that may be imagined, till such time as other phenomena occur, by which they may either be made more accurate, or liable to exceptions.

This rule we must follow, that the argument of induction may not be evaded by hypotheses.

In addition to everything thus far considered, the *Principia* unveiled a multitude of other discoveries, theories, and definitions critical to modern science. In the *Principia,* Newton was the first to clearly differentiate between the concepts of mass and weight. He defined, for the first time, centripetal force. Newton also laid the groundwork for the new sciences of rational dynamics and fluid dynamics. And, he successfully used his law of gravitation to derive the orbits of known comets, showing that some comets orbited on parabolic paths, others on hyperbolic paths, and yet others orbited the sun on elliptical paths. It was this last class of comets that reappeared periodically from the viewpoint of the earth, much like Halley's Comet. Any one of these groundbreaking discoveries would make a book worthy of admiration; combine them all into one book and the result is the most important publication in the history of science.

The *Opticks*

The *Principia* is a masterful example of deductive science. It is mathematical physics at its best — in fact, in the *Principia* Newton essentially invented mathematical physics. On the other hand, Newton's other major work, the *Opticks* (1704), is a classic example of inductive, or experimental, science. The fact that two such important works representing philosophies on the opposite ends of the scientific spectrum arose from the pen of one man is evidence of Newton's brilliance.

In the *Opticks*, Newton wrote about his experiments that attempted to settle many questions concerning the nature of light. Newton completed many of his experiments, along with the deductions he made about their results, decades earlier. In fact, his first published paper concerned the nature of light. Unfortunately, his conclusions were antithetical to many commonly held assumptions, and the young Newton came under attack from many, not the least of whom was Robert Hooke. Newton's response was to withdraw from the public eye and refuse to publish any more results for many years to come.

When Newton did finally publish the *Opticks*, the book established Newton's corpuscular theory of light. The corpuscular theory maintained that light was composed of tiny particles and that the behavior of light can best be understood by attempting to understand the behavior of these particles. Many of Newton's opponents supported the view of Christiaan Huygens, who maintained that light behaved as a wave.*

The traditional Aristotelian theory of white, or natural, light was that it could be modified to produce colors. This is exactly what happened to white light, Aristotelians maintained, as it passed through a prism to produce the colors of the spectrum — the prism modified the white light to produce the colors. Through a series of carefully constructed experiments, Newton showed that white light was a heterogeneous mixture of the various colors and these colors were themselves primary and unchanging. In one experiment Newton passed white light through a prism, separating it into its component colors. To demonstrate that each resulting color was primary, Newton allowed the individual color to pass through another prism with no change noted. In other words, if pure blue entered a prism, pure blue left the other side of the prism. Blue is a primary color and cannot be broken into components like white light.

Another experiment involved separating white light into its components — the colors of the spectrum — and then reconstituting the resulting spectrum into white light using a lens:

PROP. I. Theor. I.

The Phenomena of Colors in refracted or reflected light are not caused by new modifications of the light variously impressed, according to the various terminations of the light and shadow.
    The Proof by Experiments.

[Newton presents several experiments to support his proposition. Following is a typical example.]

---

*The nature of light continued to be a major debate among scientists into the twentieth century, when physicists agreed on the dual nature of light — it can simultaneously be considered to have properties of both particles and waves.

Exper. 2. The Sun's light let into a dark chamber through the round hole F, half an inch wide, passed first through the prism ABC placed at the hole, and then through a lens PT something more than four inches broad, and about eight feet distant from the prism, and thence converged to O the focus of the lens distant from it about three feet, and there fell upon a white paper DE. If that paper was perpendicular to that light incident upon it, as tis represented in the posture DE, all the colors upon it at O appeared white. But if the paper being turned about an axis parallel to the prism, became very much inclined to the light, as tis represented in the positions *de* and δε; the same light in the one case appeared yellow and red, in the other blue. Here one and the same part of the light in one and the same place, according to the various inclinations of the paper, appeared in one case white, in another yellow or red, in a third blue, whilst the confine of light and shadow, and the refractions of the prism in all these cases remained the same.

The *Opticks* contained experiment after experiment such as these demonstrating the nature and characteristics of light.

Newton closed the *Opticks* with a set of Queries. In spite of his very vocal aversion to hypotheses, these Queries were just that — a set of hypotheses outlining many of Newton's fundamental ideas concerning the universe and its natural laws. Originally sixteen in number, Newton added to these Queries in subsequent editions until a total of 31 Queries appeared. Many of these Queries were attempts to support a theory of Newton or to discredit opposing views. For instance, Query 28 was a pointed attack on Descartes' theory of the plenum:

Are not all hypotheses erroneous, in which light is supposed to consist in pression or motion, propagated through a fluid medium? For in all these hypotheses the phenomena of light have been hitherto explained by supposing that they arise from new modifications of the rays; which is an erroneous supposition.

Newton proceeded to argue in detail that errors arise when explaining the nature of light through the filter of constant contact in a fluid medium — Descartes' plenum.

Although today the *Opticks* is considered only Newton's second most important work, in the eighteenth century it was more popular than the *Principia*, going through more editions and selling many more copies. The *Principia* was also much less accessible to the reading public because of its mathematical nature, making it difficult to comprehend for all except a selected few capable of understanding the mathematics of the book. The *Opticks*, on the other hand, was comprised of experiments that were understandable and repeatable (for the most part) by a wide range of people. Either book would place the author among the most important scientists in history. That Newton was the author of both books places him above the rest.

The Discovery of Calculus

Isaac Newton is most famous for his discovery of the laws of motion and for the universal law of gravitation. However, one of his most long-lasting contributions to science was actually not in any field of science itself, but rather in mathematics. Calculus is crucial for the development of modern mathematics, but, more than that, the problems to which calculus has been applied range from physics to economics and almost any other discipline one can imagine. The story of the development of calculus is full of intrigue.

Who invented calculus? Most historians today would say Isaac Newton and the German natural philosopher Gottfried Leibniz invented calculus independently in the late seventeenth century. But the real story is more complicated (and much more interesting). First of all, calculus was not invented suddenly at any one point in history. We can trace the evolution of calculus all the way back to the Greek mathematician and engineer Archimedes (287 B.C.— 212 B.C.). Archimedes' methods for finding the areas enclosed by various geometric figures closely resemble the modern techniques of integral calculus. Jumping forward to the seventeenth century, we find in the work of such notable scientists as Kepler, Galileo, and Torricelli the nascent roots of calculus. In addition, other mathematicians like Fermat, Pascal, Wallis, and even Barrow (Newton's teacher and predecessor in the Lucasian chair) made significant contributions to the disciplines known today as integral and differential calculus.

So why, if all of these men and many others worked in areas now considered calculus, are Newton and Leibniz considered the inventors of the discipline? The primary reason is that it was Newton and Leibniz who were able to see the connection between the two branches, integral and differential calculus. In fact, they realized that these two seemingly distinct areas of mathematics were actually inverses of one another, a realization that made the discipline of calculus a reality. Sharing credit between Newton and Leibniz is easy to do today, but that was decidedly *not* the case in the seventeenth century. In fact, the question of which of these two men could claim priority of the discovery was both hotly disputed and lasted long after each had died.

Isaac Newton developed his version of calculus (he actually called it the method of fluxions) during his years of peak productivity when the plague sent him home from Cambridge. Although he wrote a short tract on fluxions in 1666, Newton chose not to publish and, in fact, published nothing on calculus for many decades. Revealed to only a few confidants, Newton's calculus remained essentially a secret for many years.

In the meantime, a young Leibniz became interested in mathematics, and much like Newton absorbed the most advanced mathematics available to him in a relatively short time. By the mid–1680s, Leibniz began publishing his findings on calculus and was hailed as the inventor of a powerful and important new tool for the application of mathematics to the natural world. Slowly, but surely, Newton and his supporters began to question the originality of the young German's work. Had he studied one of the manuscripts circulating in England that outlined Newton's method of fluxions? Or was it Newton, as the followers of Leibniz rebutted, who was now trying to claim credit for a discovery that wasn't his? The debate raged on, as most continental mathematicians sided with Leibniz, and — as would be expected — most British mathematicians supported Newton.

One reason Leibniz received (and continues to receive) support as the discoverer of calculus is that he taught his calculus to a number of talented students who expanded the techniques and spread them around the world. One of these students, John Bernoulli (brother of Jacob Bernoulli, another very important name in early calculus), decided to challenge the mathematical community with a difficult mathematical problem. Called the brachistichrone problem, no solutions were received for several months. Bernoulli and Leibniz, knowing that the problem could only be solved by someone intimately familiar with calculus, sent the challenge to Newton confident that he would not be successful either. Not only did Newton solve the problem, but, according to an account from his nephew, solved it in one night.

Today historians generally credit Newton and Leibniz as co–inventors of calculus. Although it is apparent that Newton conceived of calculus many years before Leibniz, it seems almost as certain that the German mathematician did develop his version of calculus independently. In fact, it is Leibniz's version, with its simpler and more intuitive notation, that survives today.

The "Other" Newton

Isaac Newton is remembered today as one of the most important scientists in history — an icon of the scientific revolution. But there was a side to Newton that few people know, in fact a side that was unknown to history until centuries after his death. A large chest, full of unpublished notes and other writings, was left to Newton's niece upon his death. This chest was passed on to descendents of Newton until the papers were auctioned in 1936 and dispersed. The British economist John Maynard Keynes began collecting what papers he could. Eventually, these papers revealed to the world a Newtonian mind quite different than that of legend.

During a large part of his life, Isaac Newton was disinterested in mathematics, physics, optics, and the like. Instead, Newton concentrated his efforts on theology and alchemy, producing (but not publishing) voluminous notes and papers on both subjects. In fact, modern historians argue that alchemy was Newton's consuming passion and that Newton should even be considered an alchemist first, as he probably considered himself. Newton envisioned himself as a continuation of a long line of alchemists, tracing his alchemical roots to ancient cultures and the Hermetic tradition. Newton's study of alchemy seemed to have taken on the same goals as those of traditional alchemy —finding the secrets of transmutation, particularly the transmutation of base elements into gold.

Why did Newton never publish his alchemical papers? Perhaps it was because he was never successful in his experiments; in particular, he never accomplished the task of transmutation of elements. Or perhaps Newton, following in the tradition of his fellow alchemists, preferred to maintain a veil of secrecy to protect any advances he thought he made from the prying eyes of others. Most likely, Newton decided to keep his alchemical studies secret because of the expected criticism and public backlash that would surely come if it became known that the iconic Newton was a secret alchemist.

In addition to his work in alchemy, the Newton papers rediscovered in the twentieth century revealed a man obsessed with biblical prophecy and symbolism. He studied and wrote extensively on the entire Bible, spending a considerable amount of time and energy on the prophecies of the Old Testament, in particular the Book of Daniel, and on the mathematical symbolisms found in the design and description of Solomon's Temple.

Newton was a Stoic, which meant he believed that everything in nature was instilled with a breath of life (*pneuma*) by God. Newton believed that Descartes' mechanical philosophy left no room for an active God. He also believed that one could find a unity in natural law by combining mathematics, natural philosophy, experimentalism, theology, and alchemy.

Everything Newton studied and wrote about was informed by his deep-seated religious beliefs and his interest in the occult of alchemy. In fact, we tend to place Newton the alchemist, or Newton the biblical scholar, in a box separate from Newton the scientist. This, however, is not the way Newton understood his studies. For Newton, everything he studied — mathematics, natural philosophy, alchemy, the Bible — were parts of one quest: the pursuit of the secrets that explained nature and the workings of the universe.

Newton's Impact

The science of Isaac Newton inspired generations of scientists and brought a new outlook to the scientific world. Rational mechanics — the application of mathematical physics to mechanical problems — played an important role in the upcoming Industrial Revolution. Men such as James Watt (who patented the first practical steam engine) became what were essentially the first mechanical engineers, combining scientific theory with mechanical know-how to innovate and build the machines that would create a new world.

The work of Isaac Newton was, of course, revered in England. On the other hand, much of his work was slow to be accepted in continental Europe, especially in France. Newtonianian philosophy was often in direct conflict with Cartesian philosophy. For instance, calculations based on Newton's law of gravitation predicted that the Earth was not a perfect sphere, but rather was flattened slightly at the poles and bulged slightly around the middle, somewhat like an onion. Proponents of Descartes' physics, on the other hand, claimed that the earth was slightly elongated at the poles. Ironically, it was a Frenchman, Pierre Louis Maupertuis, who proved that the Newtonian view of the shape of the earth was the correct one. Maupertuis led an expedition to Lapland, an area in the far northern reaches of Scandinavia, where he measured a degree of latitude. By comparing this measurement to similar measurements made at other locations, he showed that the Earth was indeed slightly flattened at the poles as Newton's followers had predicted. The implications for the Newtonian-Cartesian debate were huge, and Newtonian science began to take over in the minds of Europeans.

One of the most interesting consequences of Newton's work was its influence on a wide variety of human activities outside the realm of science. Newton's success in bringing order to the chaotic world of natural philosophy inspired a host of imitators from all walks of intellectual life. Newtonian science was a model of order, harmony and consistency in the universe, and these ideas were adapted to non-scientific social problems. Enlightenment philosophers, influenced by Newtonian methods and philosophy, looked for the first time upon science as the key to human progress. Newtonian principles were applied to physiology, politics, religion, morality, and social institutions. These ideas spread from England all over Europe, indeed all over the world. Taken to its extreme, Newtonianism was proposed as a cure for all of mankind's ills. The nineteenth century French philosopher Henri Saint-Simon went so far as to propose the establishment of a formal religion of Newton where scientists served as priests and redemption came through science instead of Christ.

Other, less extreme, versions of Newtonianism were espoused by well-respected leaders in various disciplines. For instance, the French writer and philosopher Voltaire was an avid Newtonian. Bishop Berkeley wrote that there is a principle of attraction in the minds of men analogous to gravitational attraction. This social force draws men together into social and political organizations. David Hume had similar ideas. He sought to produce a science of moral behavior that paralleled Newton's natural philosophy. Economist Adam Smith — the father of capitalism — associated the natural price of commodities with laws of gravity. He wrote that prices are always gravitating towards this natural price. Even in America, Newtonian principles were embraced for what they said about politics by Franklin, Jefferson, Adams and others. James Wilson, a delegate to the Constitutional Convention, wrote that political bodies have an inherent inertia, using the scientific term to describe a political body.

Newtonian principles even helped shape theological philosophies. The minister Richard Bentley argued that Newton's regular laws proved the existence of God. A new concept of religion, one based on rational thought rather than revealed knowledge, began to emerge. An argument called the argument from design appeared that said a designed system such as our universe implied the existence of a designer. In its ultimate manifestation, a distinctly un–Newton like religion emerged as deists claimed that a creator was responsible for the world as we see it but was not involved in the day-to-day operation of the universe.

The second half of the eighteenth century found Europe entering into a period of enlightenment. This movement, primarily led by the French intellectuals known as *philosophes* such as Voltaire and Rousseau, appealed to many Newtonian principles. The leaders of the Enlightenment all agreed on some basic principles: the importance of reason over ancient authority and revealed knowledge; a belief in man's ability to understand nature through reason and experimentation; a belief in progress of humankind; a suspicion of nationalism; and a fundamental belief in service to mankind. In addition, the Enlightenment thinkers believed in the application of scientific principles, especially Newtonian principles, to the social and political sciences and that a new science of man could rid the world of injustice, poverty and other ills.

By the second half of the eighteenth century, Newtonianism had become a new paradigm representing everything that modern science and modern thinking could do to benefit humankind. Newton had, in the opinion of the Enlightenment natural philosophers, finished what Copernicus had started. Even though the Newtonian universe was questioned and in some respects superseded by Einstein and other twentieth century scientists, Newtonian physics still serves to explain our world in a very accurate way. But more importantly, Newton changed society by changing the way thinkers in all disciplines approached their work.

PRIMARY SOURCES

Newton, Isaac. *The Mathematical Principles of Natural Philosophy*, translated by Andrew Motte, 1729.
_____. *The Opticks*.

OTHER SOURCES

Aughton, Peter. *Newton's Apple: Isaac Newton and the English Scientific Renaissance*. London: Weidenfeld & Nicolson, 2003.

Berlinski, David. *Newton's Gift: How Sir Isaac Newton Unlocked the System of the World*. New York: Free Press, 2000.

Buchwald, Jed Z., and Bernard I. Cohen. *Isaac Newton's Natural Philosophy*. Cambridge: MIT Press, 2001.

Christianson, Gale E. *Isaac Newton*. Oxford and New York: Oxford University Press, 2005.

Dobbs, Betty Jo Teeter, and Margaret C. Jacob. *Newton and the Culture of Newtonianism*. Amherst, NY: Humanity Books, 1995.

Feingold, Mordechai. *The Newtonian Moment: Isaac Newton and the Making of Modern Culture*. New York: New York Public Library: Oxford University Press, 2004.

Force, James E., and Sarah Hutton. *Newton and Newtonianism: New Studies*. Dordrecht and Boston: Kluwer Academic, 2004.

Force, James E., and Richard H. Popkin. *Essays on the Context, Nature, and Influence of Isaac Newton's Theology*. Dordrecht and Boston: Kluwer Academic, 1990.

Gleick, James. *Isaac Newton*. New York: Pantheon Books, 2003.

Hall, A. Rupert. *Isaac Newton, Adventurer in Thought*. Oxford and Cambridge, MA: Blackwell, 1992.

Westfall, Richard S. *Isaac Newton*. Oxford and New York: Oxford University Press, 2007.

_____. *The Life of Isaac Newton*. Cambridge and New York: Cambridge University Press, 1993.

White, Michael. *Isaac Newton: The Last Sorcerer*. Reading, MA: Addison-Wesley, 1997.

_____. *Never at Rest: A Biography of Isaac Newton*. Cambridge and New York: Cambridge University Press, 1980.

CHAPTER 8

Lavoisier, Dalton, and the Birth of Modern Chemistry

We must trust to nothing but facts: These are presented to us by nature, and cannot deceive. We ought, in every instance, to submit our reasoning to the test of experiment, and never to search for truth but by the natural road of experiment and observation.—Antoine Lavoisier, *An Elementary Treatise on Chemistry* (1789)

Therefore we may conclude that the ultimate particles of all homogeneous bodies are perfectly alike in weight, figure, &c. In other words, every particle of water is like every other particle of water, every particle of hydrogen is like every other particle of hydrogen, &c.— John Dalton, *A New System of Chemical Philosophy* (1808)

Before the eighteenth century, the science of chemistry did not exist in any modern sense. The concept of an element continued to be dominated by the Aristotelian definition of four principles that composed the terrestrial sphere: earth, water, air, and fire. All substances were composed of a different combination of these four principles. Although alchemists employed the tools of chemistry in the laboratory and traced their lineage to ancient times, the ultimate goal of transmuting base metals into gold had strong undercurrents of magic and the occult. A movement called iatrochemistry, introduced by the Swiss philosopher Paracelsus in the sixteenth century, combined ideas from pharmacy, medicine, magic, astrology, alchemy, and philosophy to advocate for a kind of holistic medicine to be practiced by physicians. Even the mechanical chemistry movement, led by Robert Boyle and intent on ridding chemistry of the supernatural, was at its core deductive and attempted to explain chemistry within the ideological framework of the mechanical philosophy. So by the mid–eighteenth century, with modern astronomy firmly established by widespread acceptance of the heliocentric model proposed by Copernicus, and modern physics originating in the work of Galileo and Newton, chemistry remained, at best, a pseudo-science. The work of a new generation of chemists—perhaps the first generation of chemists—began to bring modern chemistry to the world.

Antoine Lavoisier

Antoine Lavoisier (1743–1794), was born into a wealthy, aristocratic family in Paris. Young Lavoisier was afforded the finest education, training to be a lawyer like his father. However, he became fascinated with science, particularly geology and chemistry, and began pursuing studies in these areas. Lavoisier was elected to the very prestigious French Academy of Sciences at the age of 25. In the ensuing three years, Lavoisier made two choices that greatly affected his life and work. First, he became a member of the farmers general, an organization charged with collecting various taxes for the French government. This was a lucrative position whose holders were generally despised due to widespread corruption. Lavoisier, however, appears to have approached his duties as tax collector honestly. The other action taken by Lavoisier was to marry the daughter of one of his partners in the farmers general. His young bride (only 13 years old at the time of their marriage) became his partner in science, learning English in order to translate the works of the British philosophers for her husband and playing hostess to a myriad of scientists who regularly visited the Lavoisier home.

Lavoisier remained active in politics throughout his life. He proposed and worked on many political and social reforms in the years leading up to the French Revolution. After the revolution, Lavoisier was an integral part of the commission that established the metric system in France, the first country in the world to do so. Unfortunately for Lavoisier, he made enemies who became leaders during the Reign of Terror and when all of the former tax farmers were rounded up for trial, none were spared. Lavoisier's life and career were cut short by the guillotine in 1794. During his fifty years, however, Lavoisier led another kind of revolution — one that would bring chemistry out of the shadows of alchemy and scholasticism and into the light of modern science.

The Phlogiston Debate

Lavoisier is often referred to as the "father of chemistry" for his many contributions to the science. In fact, Lavoisier's work is probably the first that would be recognized by a modern student as chemistry. One of the most important contributions made by Lavoisier to the new science of chemistry was his theory of combustion that rivaled, and eventually superseded, the theory held by many natural philosophers of the time.

Natural philosophers had always searched for an explanation of combustion — Why did things burn? More particularly, why did some objects burn readily, others not so readily, and still others not at all? Many theories developed to explain the phenomenon of combustion. During the eighteenth century, one particular theory —first stated by the German physician J.J. Becher (1635–1682) but popularized by another German physician, Georg Stahl (1660–1734) — had become predominant. Stahl argued that all combustible materials contained a substance he called phlogiston, and the combustion process released this phlogiston into the air. An entire theory of combustion, and even a naming system for different gases, or airs, evolved from Stahl's phlogiston proposal. For instance, when the air surrounding a burning object in an enclosed space became saturated with the phlogiston being released, the combustion process stopped. The surrounding air had become phlogisticated. This meant, of course, that a type of air that fully supported combustion — what we call oxygen today — was readily able to absorb large amounts of phlogiston. Thus, oxygen was at one time known as dephlogisticated air. Envision this kind of air as a dry sponge, ready to absorb the maximum amount of liquid. As the sponge becomes saturated, its ability to absorb more liquid eventually stops altogether. Although there were some obvious flaws in the phlogiston theory, its ability to explain many of the observations surrounding combustion and other chemical reactions meant that it was widely accepted for some time by natural philosophers.

The eventual realization that combustion did not result from a release of phlogiston (or any other substance, for that matter) was the result of one discovery made by a man who, ironically, remained committed to the phlogiston theory. Joseph Priestley was an English theologian and political commentator who became interested in science, especially electricity and chemistry. Through a series of experiments, Priestley concluded that, contrary to Aristotle and thousands of years of tradition, air was not an element, but rather composed of different kinds of airs. In particular, Priestley isolated one type of air that supported combustion to a much greater extent than common (atmospheric) air. Because of its ability to absorb a large amount of phlogiston, Priestley named his discovery dephlogisticated air. Lavoisier, however, soon renamed Priestley's discovery oxygen.

The discovery of oxygen is one of the most interesting episodes in the history of science. The roots of the discovery of this critical part of air are so intertwined that even today the question "Who discovered oxygen?" does not have a clear answer. The story begins with Joseph Priestley. Priestley was born in humble circumstances to a cloth maker and his devout wife. He studied for the ministry, a profession which he embraced for the remainder of his life. Priestley, however, was a Dissenter, and therefore was shunned by the church-supported universities such as Cambridge and Oxford. Eventually, his religious views and his support for the revolution in America made it imperative that Priestley immigrate to the new world, where he settled in Pennsylvania with his family. Priestley was hailed by the American people for his scientific work and support of their revolution. His scientific life in American, however, consisted primarily of defending the phlogiston theory against the encroaching oxygen theory put forth by Lavoisier and spreading rapidly through the scientific community.

During the course of one of his experiments, Priestley produced an air that had all of the characteristics of atmospheric air, but intensified. It supported a flame to a greater degree, and, Priestley found, it supported respiration in mice much longer than normal air, as he related in *Experiments and Observations on Different Kinds of Air*:

> The flame of the candle, besides being larger, burned with more splendor and heat ... and a piece of red-hot wood sparkled in it ... and it consumed very fast.... On the eighth of this month, I procured a mouse, and put it into a glass vessel, containing two ounce-measures of the [dephlogisticated] air.... Had it been common air, a full-grown mouse, as it was, would have lived in it about a quarter of an hour. In this air, however, my mouse lived a full half hour....

Priestley continued his experiments, producing purer forms of his newly discovered air, and finding that mice could live up to four times longer in this environment — a surprising discovery. In fact, Priestley's experiments continually presented surprises; so much so, that he felt the need to apologize to his readers for dwelling on the subject: "I wish my reader be not quite tired with the frequent repetition of the word surprise, and others of similar import; but I must go on in that style a little longer."

To understand why Priestley expressed such surprise over his findings, it must be remembered that he and others were in the process of overturning a fundamental precept of natural philosophy — a precept that had been essentially unquestioned for 2000 years. Air, it seems, is *not* an element, as Aristotle had claimed. Priestley explained to the reader the hold that this idea had on him and others:

> There are, I believe, very few maxims in philosophy that have laid firmer hold upon the mind than that of air, meaning atmospheric air (free from various foreign matters, which were always supposed to be dissolved, and intermixed with it) is a simple, elementary substance, indestructible, and unalterable, as least as much so is water supposed to be.

Priestley did not expect his experiments to take him where they did; thus, the continual series of surprises. Priestley echoed the feelings of scientists that came before and those that would come after, when he admits that these preconceived notions in effect put blinders on him and others:

> For my own part, I will frankly acknowledge, that, at the commencement of the experiments recited in this section, I was so far from having formed any hypothesis that led to the discoveries I made in pursuing them, that they would have appeared very improbably to me had I been told of them; and when the decisive facts did at length obtrude themselves upon my notice, it was very slowly, and with great hesitation, that I yielded to the evidence of my senses. And yet, when I reconsider the matter, and compare my last discoveries relating to the constitution of the atmosphere with the first, I see the closest and the easiest connection in the world between them, so as to wonder that I should not have been led immediately from the one to the other. That this was not the case, I attribute to the force of prejudice, which, unknown to ourselves, biases not only our judgments, properly so called, but even the perceptions of our senses; for we may take a maxim so strongly for granted, that the plainest evidence of sense will not entirely change, and often hardly modify our persuasions; and the more ingenious a man is, the more effectually he is entangled in his errors; his ingenuity only helping him to deceive himself, by evading the force of truth.

So, in spite of his own prejudices and preconceived notions, Priestley wrote that he was "soon satisfied that atmospheric air is not an unalterable thing."

Joseph Priestley is usually accorded the title of "discoverer of oxygen." However, this title might be problematic. Several chemists had isolated oxygen years before Priestley, including Robert Boyle. None of them, however, understood that they had separated an element from air. About two years before Priestley's discovery, a Swedish chemist by the name of Carl Wilhelm Scheele produced oxygen, calling it fire air. However, Scheele did not publish his results for many years. Lavoisier also laid claim to the discovery, although it seems apparent that he knew of Priestley's experiment and therefore was not original in his work. However, Lavoisier was the first to prove that this new air was in fact an element, and he was the first to explain the correct role of oxygen in the combustion process, to the discredit of the phlogiston theory.

Lavoisier heard about Priestley's results and, after conducting experiments concerning the nature of this new air himself, concluded that the substance isolated by Priestley was actually *absorbed* in the combustion process. Notice that this theory was essentially the reverse of the phlogiston theory, which stated a substance was *released* during combustion. Lavoisier renamed Priestley's discovery *oxygen*, meaning acid-maker. Not only had Lavoisier put forth a new theory of combustion, he also began his program for changing the way chemical substances were named. Although the phlogiston theory was not given up easily by many chemists (Priestley and Lavoisier clashed over the theory of combustion for the rest of their respective lives), eventually the oxygen theory of combustion triumphed and formed one of the important foundations for modern chemistry. So although Joseph Priestley usually gets the credit, the discovery of oxygen and its true nature was the work of a series of experiments conducted by several chemists.

An Elementary Treatise on Chemistry

In the preface to his book, *An Elementary Treatise on Chemistry*, Lavoisier related to the reader his purpose in pursuing a new naming system for chemical substances as well as the new directions his work took:

> When I began the following work, my only object was to extend and explain more fully the memoir which I read at the public meeting of the academy of sciences in the month of April 1787, on the necessity of reforming and completing the nomenclature of Chemistry.... Thus, while I thought myself employed only in forming a nomenclature, and while I proposed to myself nothing more than to improve the chemical language, my work transformed itself by degrees, without my being able to prevent it, into a treatise upon the Elements of Chemistry.
>
> The impossibility of separating the nomenclature of a science from the science itself, is owing to this, that every branch of physical science must consist of three things; the series of facts which are the objects of the science, the ideas which represent these facts, and the words by which these ideas are expressed.
>
> We must trust to nothing but facts: These are presented to us by nature, and cannot deceive. We ought, in every instance, to submit our reasoning to the test of experiment, and never to search for truth but by the natural road of experiment and observation.

Lavoisier continued by noting that the age-old Aristotelian assumption of four elements was crumbling quickly due to the discoveries in the laboratories around Europe:

> It will, no doubt, be a matter of surprise, that in a treatise upon the elements of chemistry, there should be no chapter on the constituent and elementary parts of matter; but I shall take occasion, in this place, to remark, that the fondness for reducing all the bodies in nature to three or four elements, proceeds from a prejudice which has descended to us from the Greek Philosophers. The notion of four elements, which, by the variety of their proportions, compose all the known substances in nature, is a mere hypothesis, assumed long before the first principles of experimental philosophy or of chemistry had any existence. In those days, without possessing facts, they framed systems; while we, who have collected facts, seem determined to reject them, when they do not agree with our prejudices. The authority of these fathers of human philosophy still carry great weight, and there is reason to fear that it will even bear hard upon generations yet to come.
>
> All that can be said upon the number and nature of elements is, in my opinion, confined to discussions entirely of a metaphysical nature. The subject only furnishes us with indefinite problems, which may be solved in a thousand different ways, not one of which, in all proba-

bility, is consistent with nature. I shall therefore only add upon this subject, that if, by the term *elements*, we mean to express those simple and indivisible atoms of which matter is composed, it is extremely probable we know nothing at all about them.

An Elementary Treatise on Chemistry is an introductory text for the new chemistry, but at the same time a careful explanation of Lavoisier's own chemical experiments and a thorough discussion of the new naming system he developed for chemical elements and compounds. For instance, in the chapter in which Lavoisier discussed the composition of atmospheric air, he described his experiments in separating and measuring the relative quantities of the two primary components of air, as well as the new naming system he proposed for identifying these elements. After isolating one part of air, Lavoisier related this experiment to determine its combustibility:

> Having filled a bell-glass of about six pints measure, with pure air, or the highly respirable part of air, I transported this jar by means of a very flat vessel, into a quicksilver bath in the basin BC, and I took care to render the surface of the mercury perfectly dry both within and without the jar with blotting paper. I then provided a small capsule of china-ware D, very flat and open, in which I placed some small pieces of iron, turned spirally, and arranged in such a way as seemed most favorable for the combustion being communicated to every part. To the end of one of these pieces of iron was fixed a small morsel of tinder, to which was added about the sixteenth part of a grain of phosphorus, and, by raising the bell-glass a little, the china capsule, with its contents, were introduced into the pure air. I know that, by this means, some common air must mix with the pure air in the glass; but this, when it is done dexterously, is so very trifling, as not to injure the success of the experiment. This being done, a part of the air is sucked out from the bell-glass, by means of a siphon GHI, so as to raise the mercury within the glass to EF; and, to prevent the mercury from getting into the siphon, a small piece of paper is twisted round its extremity. In sucking out the air, if the motion of the lungs only be used, we cannot make the mercury rise above an inch or an inch and a half; but, by properly using the muscles of the mouth, we can, without difficulty, cause it to rise six or seven inches.
>
> I next took an iron wire, properly bent for the purpose, and making it red hot in the fire, passed it through the mercury into the receiver, and brought it in contact with the small piece of phosphorus attached to the tinder. The phosphorus instantly takes fire, which communicates to the tinder, and from that to the iron. When the pieces have been properly arranged, the whole iron burns, even to the last particle, throwing out a white brilliant light similar to that of Chinese fireworks. The great heat produced by this combustion melts the iron into round globules of different sizes, most of which fall into the China cup; but some are thrown out of it, and swim upon the surface of the mercury. At the beginning of the combustion, there is a slight augmentation in the volume of the air in the bell-glass, from the dilatation caused by the heat; but, presently afterwards, a rapid diminution of the air takes place, and the mercury rises in the glass; insomuch that, when the quantity of iron is sufficient, and the air operated upon is very pure, almost the whole air employed is absorbed.

This sort of careful description of his experiments is typical of the writing in *An Elementary Treatise on Chemistry*. Not only was Lavoisier creating a new naming system for chemical substances, he was establishing the new science of chemistry on solid experimental foundations.

Lavoisier proceeded to explain his reasons for naming the two primary components of air. He picked the name oxygen, from the Greek word meaning acid-former, for the combustible constituent of air.

We have already seen, that the atmospheric air is composed of two gasses, or aëriform fluids, one of which is capable, by respiration, of contributing to animal life, and in which metals are calcinable, and combustible bodies may burn; the other, on the contrary, is endowed with directly opposite qualities; it cannot be breathed by animals, neither will it admit of the combustion of inflammable bodies, nor of the calcination of metals. We have given to the base of the former, or respirable portion of the air, the name of *oxygen*, from οξυς *acidum*, and γεινομας, *gignor*; because, in reality, one of the most general properties of this base is to form acids, by combining with many different substances. The union of this base with caloric we term *oxygen gas*, which is the same with what was formerly called *pure*, or *vital air*. The weight of this gas, at the temperature of 10°, and under a pressure equal to 28 inches of the barometer, is half a grain for each cubical inch, or one ounce and a half to each cubical foot.

The name of the other primary component of air is a bit less certain according to Lavoisier's account:

The chemical properties of the noxious portion of atmospheric air being hitherto but little known, we have been satisfied to derive the name of its base from its known quality of killing such animals as are forced to breathe it, giving it the name of *azote*, from the Greek primitive particle α and ξαη, vita; hence the name of the noxious part of atmospheric air is *azotic gas*; the weight of which, in the same temperature, and under the same pressure, is 1 *oz.* 2 *gros.* and 48 *grs.* to the cubical foot, or 0.4444 of a grain to the cubical inch. We cannot deny that this name appears somewhat extraordinary; but this must be the case with all new terms, which cannot be expected to become familiar until they have been some time in use. We long endeavored to find a more proper designation without success; it was at first proposed to call it *alkaligen gas*, as, from the experiments of Mr. Berthollet, it appears to enter into the composition of ammoniac, or volatile alkali; but then, we have as yet no proof of its making one of the constituent elements of the other alkalies; beside, it is proved to compose a part of the nitric acid, which gives as good reason to have called it *nitrogen*. For these reasons, finding it necessary to reject any name upon systematic principles, we have considered that we run no risk of mistake in adopting the terms of *azote*, and *azotic gas*, which only express a matter of fact, or that property which it possesses, of depriving such animals as breathe it of their lives.

The name azote, meaning without life, came from the fact that the gas does not support respiration in animals. As it became apparent that this gas did support plant life, the name azote was eventually replaced with nitrogen, meaning nitre, or ash, forming.

Lavoisier's experiments with common air, his discoveries concerning the role of oxygen in the combustion process, and his new naming system for the various types of airs, were the culmination of a decades-long frenzy of work by a handful of early chemists to isolate and understand new elements (and compounds, as it turned out), especially gases. Joseph Black (1728–1799), a Scottish physician and chemist, isolated a gas in his lab with several interesting properties. For one thing, the gas did not support combustion. Neither did it support respiration in animals. Black called his discovery fixed air because it was released from limestone upon heating — therefore the gas was fixed into the mineral. Black was also surprised to find that fixed air — now known as carbon dioxide — was actually released by animals during the respiration process.

Another new air was isolated by Henry Cavendish (1731–1810), a British chemist born into a well-to-do family whose wealth provided Henry with all the support he needed to finance a life in science. A very private and eccentric man, Cavendish is credited with many important discoveries in science, from an accurate measurement of the density of the earth to several discoveries in the emerging field of electricity. As one measure of Cavendish's eccentricity, many of these discoveries were never published and only emerged after his death as fellow scientists reviewed his notes and manuscripts. One discovery made by Cavendish that he did publish was that of elemental hydrogen. Working in his well-appointed laboratory, Cavendish produced this gas that was highly combustible, which he called inflammable air.

In his "Experiments on Factitious Air," Cavendish illustrated several methods for producing inflammable air. One of these involved reacting certain metals with acid: "I know of only three metallic substances, namely zinc, iron, and tin, that generate inflammable air by solution in acids."

Cavendish also found that inflammable air could be produced by organic material. In particular, he found that "the air produced from gravy broth by putrefaction" was inflammable air and "it took fire on applying a piece of lighted paper, and went off with a gentle bounce, of much the same degree of loudness as when the phial was filled with the last mentioned quantities of inflammable air from zinc and common air." Cavendish included careful measurements of the relative weights of all of the reactants in his experiments.

The work of Priestley, Cavendish, Black, Lavoisier, and others was instrumental in disproving the long-held assumption that air was an element. Many of the same chemists were involved in another shocking discovery — water is not an element either! Once again, credit is difficult to assign. Although Priestley and Cavendish discovered that they could form water by introducing an electrical spark to a mixture of inflammable air (hydrogen) and common air, neither realized the significance of the discovery. Both men, in fact, explained their results within the context of the phlogiston theory. It would be Lavoisier who repeated the experiment and concluded that water was not an element, but rather a combination of hydrogen and oxygen. In *An Elementary Treatise on Chemistry*, Lavoisier wrote:

> Until very lately, water has always been thought a simple substance, insomuch that the older chemists considered it as an element. Such it undoubtedly was to them, as they were unable to decompose it; or, at least, since the decomposition which took place daily before their eyes was entirely unnoticed. But we mean to prove, that water is by no means a simple or elementary substance.

After describing several experiments by which he separated water into its constituent parts — oxygen and another gas — Lavoisier came to the point of naming this other gas. He settled on hydrogen, meaning water-former:

> Thus water, besides the oxygen, which is one of its elements in common with many other substances, contains another element as its constituent base or radical, and for which we must find an appropriate term. None that we could think of seemed better adapted than the word *hydrogen*, which signifies the *generative principle of water*, from ύδορ *aqua*, and γεινομας *gignor*.

Lavoisier also concluded that matter exists in three states, dependent only upon the amount of heat contained in the body. Once again, Lavoisier was not the first to make this important claim — the conclusion was first reached by the French economist Anne-Robert Jacques Turgot — but Lavoisier supported Turgot's theory and incorporated it into his new chemical science:

> All bodies in nature present themselves to us in three different states. Some are solid like stones, earth, salt, and metals. Others are fluid like water, mercury, spirits of wine; and others finally are in a third state which I shall call the state of expansion or of vapors, such as water when one heats it above the boiling point. The same body can pass successively through each of these states, and in order to make this phenomenon occur it is necessary only to combine it with a greater or lesser quantity of the matter of fire.

By the time Lavoisier published *An Elementary Treatise on Chemistry* in 1789, it was apparent that chemistry was undergoing a fundamental transformation. No longer a simplistic deductive model with only a few elements, the laboratory discoveries made by Lavoisier and his contemporaries culminated in a new structure that laid the groundwork for modern chemistry. Lavoisier's careful use of the balance in his laboratory led him to the first clear statement of the conservation of mass: in a chemical reaction the mass of the products must equal the mass of the reactants. In spite of the success of this new analytical chemistry, questions remained: "Why did elements behave chemically the way they did? What is the fundamental, physical, reality underlying chemical elements and their reactions?" To help answer such questions, an English chemist took up where the Frenchman Lavoisier left off.

John Dalton and the Revival of Atomism

John Dalton's path to chemistry was very different from that of Lavoisier. Dalton (1766–1844) was born in humble circumstances to a Quaker family in England. Without the family resources afforded Lavoisier, Dalton spent the majority of his life as a teacher at various academies and as a private tutor. As a Dissenter (like Priestley), Dalton was not eligible to teach at the major universities in England. As a young man, Dalton moved to Manchester where he joined an active group of intellectuals in the Manchester Literary and Philosophical Society. With the encouragement of his new peers in philosophy, Dalton began his life's work in science.

John Dalton is remembered as the man who brought atomism to modern chemistry. But the theory that an ultimate particle, or atom, composed all matter had ancient roots. Leucippus and Democritus, Greek philosophers of the fifth century B.C.E., proposed the existence of indefinitely small, invisible particles called *atomos*, meaning uncuttable. The two philosophers, along with another Greek named Epicurus, developed a theory of nature built around the concept of the atom. Atomic theory eventually lost favor in Greece and throughout Europe as Aristotle's natural philosophy became almost universally adopted. Aristotle rejected the possibility of empty space, or the void. The theory of atoms maintained that these small particles moved through empty space careening off one another, in direct conflict with Aristotle's teachings about a void. Atomism remained a neglected and discarded theory until the early nineteenth century, when John Dalton revived the ancient ideas of the Greek philosophers.

Dalton's experiments led him to the important conclusion that chemical combinations of elements followed laws pertaining to their relative weights. He proposed to assign hydrogen a weight of one, and gave the weights of other elements (and compounds) on a scale relative to hydrogen. Shortcomings in Dalton's experiments and interpretation of the results meant that some of his atomic weights were incorrect; yet, the idea that chemical reactions were governed by a physical property as simple as weight was of great importance. In *A New System of Chemical Philosophy*, Dalton wrote:

> Before we can apply this doctrine to find the specific heat of elastic fluids, we must first ascertain the relative weights of their ultimate particles. Assuming at present what will be proved hereafter, that if the weight of an atom of hydrogen be 1, that of oxygen will be 7, azote 5, nitrous gas 12, nitrous oxide 17, carbonic acid 19, ammoniacal gas 6, carburetted hydrogen 7, olefiant gas 6, nitric acid 19, carbonic oxide 12, sulphuretted hydrogen 16, muriatic acid 22, aqueous vapor 8, ethereal vapor 11, and alcoholic vapor 16; we shall have the specific heats of the several elastic fluids as in the following table. In order to compare them with that of water, we shall further assume the specific heat of water to that of steam as 6 to 7, or as 1 to 1.166.

Dalton claimed his theory of relative atomic weights is one of his primary purposes in composing this book:

> Now it is one great object of this work, to show the importance and advantage of ascertaining the relative weights of the ultimate particles, both of simple and compound bodies, the number of simple elementary particles which constitute one compound particle, and the number of less compound particles which enter into the formation of one more compound particle.

Dalton proceeded to give various rules for combinations of these elements to form compounds, always keeping in mind the relative weights of the elementary particles.

Dalton explained that the existence of some sort of ultimate particle was by then almost universally accepted, based on the fact that the same substance can exist in three states. However, the physical nature of these ultimate particles was not universally agreed upon. Dalton believed that this had hindered chemical research:

> There are three distinctions in the kinds of bodies, or three states, which have more especially claimed the attention of philosophical chemists; namely, those which are marked by the terms elastic fluids, liquids, and solids. A very familiar instance is exhibited to us in water, of a body, which, in certain circumstances, is capable of assuming all the three states. In steam we recognize a perfectly elastic fluid, in water, a perfect liquid, and in ice a complete solid. These observations have tacitly led to the conclusion which seems universally adopted, that all bodies of sensible magnitude, whether liquid or solid, are constituted of a vast number of extremely small particles, or atoms of matter bound together by a force of attraction, which is more or less powerful according to circumstance, and which as it endeavors to prevent their separation, is very properly called in that view, attraction of cohesion; but as it collects them from a dispersed state (as from steam into water) it is called, attraction of aggregation or more simply, affinity. Whatever names it may go by, they still signify one and the same power. It is not my design to call in question this conclusion, which appears completely satisfactory; but to show that we have hitherto made no use of it, and that the consequence of the neglect, has been a very obscure view of chemical agency, which is daily growing more so in proportion to the new lights attempted to be thrown upon it.

Dalton proposed his atomic theory, which essentially drew as a conclusion the definition of an element.

> Whether the ultimate particles of a body, such as water, are all alike, that is, of the same figure, weight, &c. is a question of some importance. From what is known, we have no reason to apprehend a diversity in these particulars: if it does exist in water, it must equally exist in the elements constituting water, namely, hydrogen and oxygen. Now it is scarcely possible to conceive how the aggregates of dissimilar particles should be so uniformly the same. If some of the particles of water were heavier than others, if a parcel of the liquid on any occasion were constituted principally of these heavier particles, it must be supposed to affect the specific gravity of the mass, a circumstance not known. Similar observations may be made on other substances. Therefore we may conclude that the ultimate particles of all homogeneous bodies are perfectly alike in weight, figure, &c. In other words, every particle of water is like every other particle of water, every particle of hydrogen is like every other particle of hydrogen, &c.

Dalton further postulated that a subtle fluid of heat surrounds the atoms, preventing the individual atoms from contacting each other.

> Besides the force of attraction, which, in one character or another, belongs universally to ponderable bodies, we find another force that is likewise universal, or acts upon all matter which comes under our cognisance, namely, a force of repulsion. This is now generally, and I think properly, ascribed to the agency of heat. An atmosphere of this subtle fluid constantly surrounds the atoms of all bodies, and prevents them from being drawn into actual contact.

The concept of a "subtle fluid" had been used by natural philosophers before Dalton to explain unseen forces. Magnetism, electricity, heat, even gravity acted upon bodies without any visible physical connections. Philosophers from William Gilbert to Isaac Newton employed the theory that an invisible, odorless, and weightless medium, a subtle fluid, permeated the universe and was responsible for these phenomena. The existence of subtle fluids remained an important theoretical construction in science until early in the twentieth century.

Dalton's experiments and atomic theory led him to a very important conclusion — mass is conserved during a chemical reaction. Dalton wrote about conservation of mass, and what a chemical reaction is and isn't, with an interesting analogy:

> Chemical analysis and synthesis go no farther than to the separation of particles one from another, and to their reunion. No new creation or destruction of matter is within the reach of chemical agency. We might as well attempt to introduce a new planet into the solar system, or to annihilate one already in existence, as to create or destroy a particle of hydrogen. All the changes we can produce, consist in separating particles that are in a state of cohesion or combination, and joining those that were previously at a distance.

One important result of a conservation of mass law such as Dalton proposed is the rejection of alchemy and its attempt to transmute metals. The work of Lavoisier, Dalton, and others in the late eighteenth and early nineteenth centuries led chemistry out of the magical and supernatural world of alchemy and into its place as a legitimate science.

The atom is, of course, an incredibly small object. This means that matter is composed of a mind-boggling number of them, as Dalton explained:

> When we attempt to conceive the number of particles in an atmosphere, it is somewhat like attempting to conceive the number of stars in the universe; we are confounded with the

thought. But if we limit the subject, by taking a given volume of any gas, we seem persuaded that, let the divisions be ever so minute, the number of particles must be finite just as in a given space of the universe, the number of stars and planets cannot be infinite.

How many elementary particles are there in a given amount of matter? Dalton never attempted to give us an answer, but others were hard at work trying to answer this question. A contemporary of Dalton, Amedeo Avogadro, claimed that there was a direct relationship between the volume of a gas and the number of atoms in the gas. Almost a century later, Jean Baptiste Jean Perrin proposed that the number of atoms in equal volumes of gases (a number that is approximately 6.02×10^{23}) be called Avogadro's Number.

Another important result deriving from Dalton's investigations into the atom was his discovery that the sum of the pressures of individual gases is equal to the total pressure of a mixture of those gases. This statement, known as Dalton's Law of Partial Pressures, forms a fundamental tenet of modern chemistry. Doors continued to open in chemistry thanks to Dalton's revival of atomic theory. One of these doors led to perhaps the most ubiquitous symbol of chemistry — a symbol found in almost every chemistry classroom and lab in every school in the world — the periodic table.

The Periodic Table

As the theory of atoms became widely accepted by chemists around the world, some began noticing a periodic relationship in the chemical and physical properties of the elements. Although various experimental and theoretical chemists contemplated the idea of periodicity, two men are credited with ultimately producing a table outlining this new and important property of the elements.

Lothar Meyer (1830–1895), a German chemist, developed his theory that elements could be grouped according to their atomic weights in such a way that the groups shared many chemical and physical properties. Unfortunately for Meyer, he withheld publication of his full theory until after a Russian contemporary emerged with the same ideas, publishing a work that more completely and more accurately described the periodic characteristics of the elements.

Dmitri Mendeleev's road to history was literally a long and winding path. Born in Siberia to a mother determined to give young Dmitri (1834–1907) every opportunity to develop his talents, mother and son set out on a treacherous journey from the hinterlands of Russia to the promise of Moscow. When Dmitri was denied entrance to the university there because of his Siberian heritage, they moved to St. Petersburg. In St. Petersburg, Dmitri was once again denied access to the university, so he matriculated at the local technical college where he began to fulfill his promise, eventually obtaining a position on the faculty at the University of St. Petersburg.

Mendeleev's research led him to the same conclusions Meyer was drawing at about the same time in Germany. In the second volume of his important *Principles of Chemistry*, Mendeleev related how he came to his theory:

> But nothing, from mushrooms to a scientific law, can be discovered without looking and trying. So I began to look about and write down the elements with their atomic weights and

typical properties, analogous elements and like atomic weights on separate cards, and this soon convinced me that the properties of elements are in periodic dependence upon their atomic weights.

Mendeleev shuffled these cards bearing the known elements until he found the correct arrangement for his periodic table. Organized in rows and columns in increasing order of atomic weight, Mendeleev found that there were certain holes in his table. The Russian had the insight to realize that these holes were meant for as yet undiscovered elements whose properties could be predicted by their placement in the table.

When Mendeleev published his periodic theory along with the table of elements, both known and not yet discovered, he was roundly criticized by his colleagues, including Meyer. First of all, in order for the properties of the elements to fit his scheme, Mendeleev was forced to rearrange a few of the elements and place them in a position contrary to where their atomic weight would have them. This ad hoc solution indicated a problem with the theory of periodicity. Secondly, Mendeleev's prediction that undiscovered elements would fit neatly into the table, and his audacity to predict very specific properties of imaginary elements, brought a great degree of ridicule upon him.

Both the man and the theory were vindicated over a period of several years when three new elements predicted by Mendeleev and his table — gallium, scandium, and germanium — were discovered. Not only did their atomic weights place them in the correct position in his periodic table, the properties of the elements were just as Mendeleev had predicted. The only problem remaining was the matter of juggling the order of some of the elements to make the periodic properties work. This problem was eventually solved, and the modern periodic table completed, by Nobel Prize-winning physicist Henry Moseley (1887–1915).

Moseley used x-ray spectroscopy to show that arranging the elements by their atomic number, rather than weight, corrected the problems faced by Mendeleev's table. This slight correction not only fixed the periodic table, it also provided a foundation for the next explosive advances in chemistry. Whereas Lavoisier had essentially created modern chemistry with a new understanding of the role of oxygen in combustion and his new definition and naming system for all elements, Dalton's revival of the atomic theory and the ultimate development of the periodic table led to a science of chemistry interested in the ultimate particles that composed matter. In the twentieth century, chemistry, combined with physics, unveiled sub-atomic particles and paved the path for a new understanding of the fundamental workings of nature.

PRIMARY SOURCES

Cavendish, Henry. "Experiments on Factitious Air." *The Scientific Papers of Henry Cavendish*, edited by Sir Edward Thorpe. Cambridge: University Press, 1921.

Dalton, John. *A New System of Chemical Philosophy*. Vol. 1. http://www.archive.org/stream/newsystemofc hemi01daltuoft/newsystemofchemi01daltuoft_djvu.txt.

Lavoisier, Antoine. *An Elementary Treatise on Chemistry*. Project Gutenberg. http://www.gutenberg.org /files/30775/30775-h/30775-h.htm.

_____. *Essays, Physical and Chemical*, translated by Thomas Henry. http://books.google.com/books ?id=HxMAAAAAQAAJ&printsec=frontcover&dq=lavoisier&hl=en#v=onepage&q&f=false.

Priestley, Joseph. "Experiments and Observations on Different Kinds of Air." *Scientific Revolutions: Primary Texts in the History of Science*, by Brian S. Baigrie. Upper Saddle River, NJ: Pearson Prentice Hall, 2004.

Other Sources

Bell, Madison Smartt. *Lavoisier in the Year One: The Birth of a New Science in an Age of Revolution.* New York: W.W. Norton, 2005.

Bensaude-Vincent, Bernadette, and F. Abbri. *Lavoisier in European Context: Negotiating a New Language for Chemistry.* Canton, MA: Science History Publications, U.S.A., 1995.

Bensaude-Vincent, Bernadette, F. Abbri, and Isabella Stengers. *A History of Chemistry.* Cambridge: Harvard University Press, 1996.

Beretta, Marco. *Imaging a Career in Science: The Iconography of Antoine Laurent Lavoisier.* Canton, MA: Science History Publications, U.S.A., 2001.

Brock, William H. *The Chemical Tree.* New York: W.W. Norton, 2000.

_____. *The Norton History of Chemistry.* New York: W.W. Norton, 1993.

Cardwell, D.S.L., ed. *John Dalton and the Progress of Science.* Manchester: Manchester University Press, 1968.

Davis, Kenneth. *The Cautionary Scientists: Priestley, Lavoisier, and the Founding of Modern Chemistry.* New York: Putnam, 1966.

Donovan, Arthur. *Antoine Lavoisier: Science, Administration, and Revolution.* New York: Cambridge University Press, 1993.

Gordin, Michael D. *A Well-Ordered Thing: Dmitri Mendeleev and the Shadow of the Periodic Table.* New York: Basic Books, 2004.

Greenaway, Frank. *John Dalton and the Atom.* Ithaca: Cornell University Press, 1966.

Guerlac, Henry. *Antoine-Laurent Lavoisier, Chemist and Revolutionary.* New York: Scribner, 1975.

_____. *Lavoisier — the Crucial Year: The Background and Origin of His First Experiments on Combustion in 1772.* Ithaca: Cornell University Press, 1961.

Holmes, Frederic Lawrence. *Lavoisier and the Chemistry of Life: An Exploration of Scientific Creativity.* Madison: University of Wisconsin Press, 1985.

Levere, Trevore Harvey. *Transforming Matter: A History of Chemistry from Alchemy to the Buckyball.* Baltimore: Johns Hopkins University Press, 2001.

McKie, Douglas. *Antoine Lavoisier: Scientist, Economist, Social Reformer.* New York: H. Schuman, 1952.

Poirier, Jean Pierre. *Lavoisier, Chemist, Biologist, Economist.* Philadelphia: University of Pennsylvania Press, 1996.

Strathern, Paul. *Mendeleyev's Dream: The Quest for the Elements.* New York: Thomas Dunne Books, 2001.

Susac, Andrew. *The Clock, the Balance, and the Guillotine; The Life of Antoine Lavoisier.* Garden City, NY: Doubleday, 1970.

Thackray, Arnold. *John Dalton: Critical Assessments of His Life and Science.* Cambridge: Harvard University Press, 1972.

Yount, Lisa. *Antoine Lavoisier: Founder of Modern Chemistry.* Springfield, NJ: Enslow, 1997.

CHAPTER 9

From Franklin to Faraday:
Developments in the Science of Electricity

I never was before engaged in any study that so totally engrossed my attention and my time as this has lately done; for what with making experiments when I can be alone, and repeating them for my Friends and Acquaintance, who, from the novelty of the thing, come continually in crowds to see them, I have, during some months past, had little leisure for anything else.—Benjamin Franklin, letter to Peter Collinson (1747)

These considerations, with their consequence, the hope of obtaining electricity from ordinary magnetism, have stimulated me at various times to investigate experimentally the inductive effect of electric currents. I lately arrived at positive results; and not only had my hopes fulfilled, but obtained a key which appeared to me to open out a full explanation of Arago's magnetic phenomena, and also to discover a new state, which may probably have great influence in some of the most important effects of electric currents.—Michael Faraday, *Experimental Researches in Electricity* (1831)

The question "Who discovered electricity?" really cannot be answered. From contact with electric eels (literally and figuratively) to static electricity to the frightening results of lightning strikes, electricity has been a part of the human experience since prehistory. As we saw in Chapter 4, William Gilbert made one of the first advances in understanding the nature of electricity when he distinguished between magnetism and static electricity. However, it was not until the work of Benjamin Franklin, Michael Faraday, and others that strides were made in understanding the nature of this mysterious phenomenon. But before scientists could conduct experiments with electricity, a method for storing electrical charge for laboratory use was required. This need was first met by a simple device of somewhat limited use, and later by one of the most important inventions for science and for modern life.

Around 1745, two inventors independently developed a device that could store a static electric charge, in essence the first capacitor. One of these inventors, the Dutch physicist Pieter van Musschenbroek, was a professor at the University of Leyden, The Netherlands — thus the device soon became known as the Leyden jar. Musschenbroek's invention of the Leyden jar was accidental. He and his assistants were attempting to store static electricity in a jar partially filled with water. An electrical charge was applied through a wire into the water, while his assistant innocently held the jar in his hand. The assistant completed the circuit and received the full discharge (and quite a jolt) of the electricity stored in the jar. Later improvements to the Leyden jar resulted in its final form — a glass jar coated inside and out by a thin metal plate, the inside plate taking the place of the water and the outside plate replacing the assistant's hand. An electrostatic charge applied to a metal rod extending from its mouth charged the metal coating on the jar. So, for the first time it became possible to store electricity for future use. A Leyden jar thusly charged could then later be employed in experiments or demonstrations. In fact, it was soon found that a series of Leyden jars connected together resulted in a powerful research tool. Benjamin Franklin noted the resemblance of the series of jars to a battery of artillery and thus applied the term battery to an electrical storage device for the first time.

Although the Leyden jar represented an important breakthrough for electrical experimenters, the device was limited in the size of charge it could hold and the length of time it was held. It was left, then, to the Italian physicist Alesandro Volta (1745–1827) to design a new device to store electricity in a more efficient and useful way. In 1800, Volta invented a device for storing electricity that came to be known as the voltaic pile, but is today recognized as the first true battery. Volta layered alternating disks of copper and zinc with cloth saturated in brine between each layer. The electrochemical reaction that ensued created a steady flow of electricity and the modern battery was born.

In a letter to Royal Society president Joseph Banks, Volta described

> the construction of an apparatus having a resemblance in its effects (that is to say, in the shock it is capable of making the arms, &c. experience) to the Leyden flask, or, rather, to an electric battery weakly charged acting incessantly, which should charge itself after each explosion; and, in a word, which should have an inexhaustible charge, a perpetual action or impulse on the electric fluid.

Volta described his new invention, along with its prime advantage over the Leyden jar — it provided a continuous electric charge:

> The apparatus to which I allude, and which will, no doubt, astonish you, is only the assemblage of a number of good conductors of different kinds arranged in a certain manner. Thirty, forty, sixty, or more pieces of copper, or rather silver, applied each to a piece of tin, or zinc, which is much better, and as many strata of water, or any other liquid which may be a better conductor, such as salt water, lye, &c. or pieces of pasteboard, skin, &c. well soaked in these liquids; such strata interposed between every pair or combination of two different metals in an alternate series, and always in the same order of these three kinds of conductors, are all that is necessary for constituting my new instrument, which, as I have said, imitates the effects of the Leyden flask, or of electric batteries, by communicating the same shock as these do ... it has no need, like these, of being previously charged by means of foreign electricity, and as it is capable of giving a shock every time it is properly touched, however often it may be.

Volta gave his invention an unusual name because of its similarity to a naturally occurring phenomenon. He noticed that the device was "much more similar ... in its form to the *natural electric organ* of the torpedo or electric eel, &c. than to the Leyden flask and electric batteries, I would wish to give the name of the *artificial electric organ*." The first battery was called, by its inventor, the *artificial electric organ* because it mimicked the properties of the electric organ found in electric eels.

Although the voltaic pile was invented too late to be of use to Benjamin Franklin, the American did make great use of the Leyden jar in his experiments. Franklin, in fact, became one of the best known scientists of his era.

Benjamin Franklin, Natural Philosopher

Benjamin Franklin is a name familiar to all. Franklin is well known for printing newspapers and *Poor Richard's Almanac*; his contributions as a founding father; one of his many inventions like the Franklin Stove or bifocals; one of his famous aphorisms "early to bed early to rise makes a man healthy, wealthy, and wise," "he that lies down with dogs shall rise up with fleas," "fish and visitors stink in three days"; or, perhaps most famously, for flying a kite in a thunderstorm. Only in this last statement do we find a slight glimpse into Franklin the scientist.

The story of Benjamin Franklin's life is one of the most cherished in American lore. Franklin (1706–1790) was born to a middle class family in Boston, where he received a brief early education that somehow led him to a life of reading, study, and self-education. Apprenticed to his brother in a Boston print shop, Ben escaped to Philadelphia at the age of 17. Working at various print shops in Philadelphia, it didn't take the talented young man long to catch the eye of the governor of Pennsylvania, who sent him to London to purchase equipment to open a newspaper in Philadelphia. After arriving in London, these plans did not materialize, leaving Franklin to support himself in a strange and frightening city by once again finding employment with a printer. By the time of his return to Philadelphia, Franklin was well-versed in the printing trade and opened his own printing shop, where he soon began publishing the newspaper *The Pennsylvania Gazette*.

Franklin's fame and wealth accelerated when he created *Poor Richard's Almanac*. With financial independence, Franklin turned his attentions to public service, serving as the postmaster for the colonies, forming the first lending library in the colonies, serving as the first president of the American Philosophical Society, and helping to establish the University of Pennsylvania, among many other civic accomplishments. Of course Franklin also became active in politics. After serving in various local positions, Franklin's reputation was such that the Pennsylvania Assembly sent him to England to intervene on their behalf against the colony proprietors, the Penn family. Franklin remained in England for almost five years. Initially working diligently to maintain the relations between the colonies and England, Franklin slowly came to the realization that independence was inevitable. His contributions to the early republic included membership on the committee to draft the Declaration of Independence, ambassador to France, delegate to the Constitutional Convention, and governor of Pennsylvania. In spite of all of his other accomplishments, late in life Franklin expressed regret that he had been born too early to witness the coming of a new era in natural philosophy. A philosopher at heart, science was perhaps Franklin's most cherished avocation. And in spite of the fact that today Franklin is best remembered for his political and civic accomplishments, during his lifetime he was hailed throughout Europe as one of the foremost scientists in the world.

Benjamin Franklin, the natural philosopher, was interested in many branches of science. He published one of the first maps accurately detailing the Gulf Stream in the Atlantic Ocean, and his observations and explanations of weather phenomena made him one of America's first meteorologists. Franklin, however, was best known (and still remembered) for his experiments with electricity and his contributions to this emerging science.

These experiments engulfed Franklin's time and thoughts for many years. In a letter to Englishman Peter Collinson, Franklin related his total immersion in electrical experimentation and in the process mentioned the public's fascination with the experiments:

> I never was before engaged in any study that so totally engrossed my attention and my time as this has lately done; for what with making experiments when I can be alone, and repeating them for my Friends and Acquaintance, who, from the novelty of the thing, come continually in crowds to see them, I have, during some months past, had little leisure for anything else.

The entertaining and playful nature of the public experiments performed by Franklin and many others masked the fact that they were serious attempts to understand the nature of this mysterious phenomenon. In one such experiment, Franklin described his results to Collinson:

> Let [person] A and [person] B stand on wax; or A on wax, and B on the floor; give one of them the electrified phial in hand; let the other take hold of the wire; there will be a small spark; but when their lips approach they will be struck and shock'd. The same if another gentleman and lady, C and D, standing also on wax, and joining hands with A and B, salute or shake hands.

In the same letter, Franklin related another imaginative experiment:

> We suspend by fine silk thread a counterfeit spider, made of a small piece of burnt cork, with legs of linen thread, and a grain or two of lead stuck in him, to give him more weight. Upon the table over which he hangs, we stick a wire upright, as high as the phial and wire, two or three inches from the spider: then we animate him, by setting the electrified phial at the same distance on the other side of him; he will immediately fly to the wire of the phial, bend his legs in touching it; then spring off, and fly to the wire in the table; thence again to the wire in the phial, playing with his legs against both, in a very entertaining manner, appearing perfectly alive to persons unacquainted. He will continue this motion an hour or more in dry weather.

Franklin described experiment after experiment in letters to Collinson and others. He used terms he had coined such as positive and negative charge, battery, conductor, and condenser.

> When a body is electrified *plus*, it will repel and electrified feather or small cork-ball. When *minus* ... it will attract them.... We made what we called an *electrical-battery*, consisting of eleven panes of large sash-glass, arm'd with thin leaden plates, pasted on each side, placed vertically, and supported at two inches distance on silk cords.

Franklin disputed the general use of the terms *electric* and *non-electric* bodies. Others had used the term *electric* body to describe a material, such as glass, "on a mistaken supposition that those called electrics, *per se*, alone contained electric matter in their substance."

Franklin argued that

the terms *electric, per se,* and *non-electric,* should be laid aside as improper: And (the only difference being this, that some bodies will conduct electric matter, and others will not) the terms *conductor* and *non-conductor* may supply their place. If any portion of electric matter is applied to a piece of conducting matter, it penetrates and flows through it, or spreads equally on its surface; if applied to a piece of non-conducting matter, it will do neither.

At the end of one letter Franklin addressed to Collinson — a letter full of experiments and observations about electricity — Franklin had a little fun (as Franklin was wont to do from time to time):

> Chagrined a little that we have been hitherto able to produce nothing in this way of use to mankind; and the hot weather coming on, when electrical experiments are not so agreeable, it is proposed to put an end to them for the season, somewhat humorously, in a party of pleasure, on the banks of *Skulkyl.* Spirits, at the same time, are to be fired by a spark sent from side to side through the river, without any other conductor than the water; an experiment which we sometime since performed, to the amazement of many. A turkey is to be killed for our dinner by the *electrical shock,* and roasted by the *electrical jack,* before a fire kindled by the *electrified bottle*: when the healths of all the famous electricians in *England, Holland, France, and Germany,* are to be drank in electrified bumpers,* under the discharge of guns from the *electrical battery.*

In other letters to Collinson, Franklin related the results of experiments he performed showing the "wonderful effect of pointed bodies, both in *drawing off* and *throwing off* the electrical fire."

Perhaps Benjamin Franklin's most enduring contribution to science was his one fluid theory of electricity. The influential chemist Charles du Fay proposed that electricity was the result of two different fluids, vitreous (or positive) and resinous (or negative). Franklin, however, disagreed with du Fay, postulating instead that there existed only one electrical fluid and that the two charges, positive and negative, could be better understood as an excess or deficiency of this fluid. Hence, electricity in Franklin's scheme was a flow of a single fluid and was conserved. Although Franklin's theory predated both the discovery of the electron and the law of conservation of charge, his one fluid theory definitely pointed scientists in the right direction for future inquiries into the nature of electricity.

Most of the practical uses for electricity to which we are so accustomed to today were many years in the future when Franklin was performing his experiments. However, Benjamin Franklin was an extremely practical man and one very important invention did come from his electrical experiments. Franklin was convinced that the true nature of lightning was electricity. The famous experiment with the kite found Franklin charging a Leyden jar by attracting lightning to a metal key and conducting the electrical energy into the awaiting jar. Although historians today dispute whether Franklin actually conducted such an experiment (Franklin never claimed in print he did — one of his correspondents, Joseph Priestley, wrote of the experiment years after the fact, and was the first written account of the experiment), Franklin did show that lightning was electricity. He proceeded to use this knowledge to make perhaps his most famous invention. At the request of a committee assigned to explore various methods to protect the gunpowder magazine at Purfleet (England) from lightning strikes, Franklin prepared a report and

---

*An electrified bumpers was a wine glass with a slight electrical charge, providing the drinker with a shock.

read it to the committee on August 27, 1772. Franklin described an experiment in this report.

> The prime conductor of an electric machine ... being supported about 10½ inches above the table by a wax-stand, and under it erected a *pointed wire* 7½ inches high, and ⅓ of an inch thick, tapering to a sharp point, and communicating with the table: when the *point* (being uppermost) is *covered* by the end of a finger, the conductor may be full charged ... but the moment the point is *uncovered* ... the prime conductor ... [is] instantly discharged and nearly emptied of its electricity. Turn the wire, its *blunt* end upwards, (which represents an unpointed bar,) and no such effect follows.

Franklin proceeded to describe an apparatus he assembled at his home, an apparatus that revealed the nature of electric experimentation in the eighteenth century — part serious science, part playful entertainment:

> In Philadelphia I had such a rod fixed to the top of my chimney, and extending about nine feet above it. From the foot of this rod, a wire (the thickness of goose-quill) came through a covered glass tube in the roof, and down through the well of the staircase; the lower end connected with the iron spear of a pump. On the staircase opposite to my chamber-door, the wire was divided; the ends separated about six inches, a little bell on each end; and between the bells a little brass ball suspended by a silk thread, to play between and strike the bells when the clouds passed with electricity in them.

And then a surprise!

> After having frequently drawn sparks and charged bottles from the bell of the upper wire, I was one night awaked by loud cracks on the staircase. Starting up and opening the door, I perceived that the brass ball, instead of vibrating as usual between the bells, was repelled and kept at a distance from both; while the fire passed sometimes in very large quick cracks as large as my finger; whereby the whole staircase was enlightened as with sun–shine, so that one might see to pick up a pin. And from the apparent quantity thus discharged, I cannot but conceive that a number of such conductors must considerably lessen that of any approaching cloud, before it comes so near as to deliver its contents in a general stroke;— an effect not to be expected from bars unpointed, if the above experiment with the blunt end of the wire is deemed pertinent to the case.

Franklin went on to describe various other experiments concerning the shape and composition of the lightning rods. He answered the most common fear concerning the lightning rod, that it attracted *more* lightning towards a building.

> It has been objected, that erecting pointed rods upon *edifices*, is to *invite* and draw lightning into *them*; and therefore dangerous. Were such rods to be erected on buildings, *without continuing the communication* quite down into the moist earth, this objection might then have weight; but when such complete conductors are made, the lightning is invited not into the building, but into the *earth*, the situation it aims at, and which it always seizes every help to obtain, even from broken partial metalline conductors.

Through careful experimentation and observation, Franklin was convinced — and successfully convinced many others — that his invention indeed worked and could save countless buildings from damage caused by lightning strikes.

When Benjamin Franklin died in 1790, he was revered as a founding father and one of the most important men of the American Revolution. He was also widely hailed — especially in Europe — as one of the outstanding scientists of the eighteenth century. His work in several fields, but especially in the relatively new discipline of electricity, represented the best of natural philosophy for his generation. He also laid the groundwork for a new generation of natural philosophers interested in pursuing an experimental and theoretical investigation of electricity.

Michael Faraday and Electromagnetic Induction

The life of Michael Faraday sounds as if it came directly from a Dickens novel. Like Benjamin Franklin, Faraday was born in humble surroundings and through his own initiative and perseverance became an international figure. Unlike Franklin, however, Michael Faraday was born in a society still rigidly class-conscious and struggled against discrimination his entire life. Even when he became an internationally renowned scientist, Faraday was often treated as an inferior due to his birth rank. In spite of these disadvantages, by the end of his life Faraday had been offered a knighthood, presidency of the Royal Society, and even burial at Westminster Abbey — all honors he turned down partially due to strict religious beliefs.

Michael Faraday (1791–1867) was born to a very poor family. His father, a blacksmith, struggled to support Michael and his siblings. Luckily for the young man, he was apprenticed to a bookbinder and found a love for knowledge through the many varieties of books he had at his disposal. Since the days of Isaac Newton, London had developed a strong tradition of public science lectures, and young Faraday took full advantage of this free education. One of these lectures, by the eminent chemist Sir Humphry Davy of the Royal Institution, marked a turning point in Faraday's life. Faraday, ever mindful of his low station in British society and his limited education, was greatly impressed by the gentleman's lecture:

> Sir H. Davy proceeded to make a few observations on the connections of science with other parts of polished and social life. Here it would be impossible for me to follow him. I should merely injure and destroy the beautiful and sublime observations that fell from his lips. He spoke in the most energetic and luminous manner of the Advancement of the Arts and Sciences. Of the connection that had always existed between them and other parts of a Nation's economy. During the whole of these observations his delivery was easy, his diction elegant, his tone good and his sentiments sublime.

Faraday worked for many years refining his own language skills and eventually overcame his lack of education and formal training. This is attested to by Auguste-Arthur de La Rive, a French physicist and friend (and admirer) of Faraday. De La Rive wrote, after a lecture given by Faraday in 1856:

> Nothing can give a notion of the charm which he imparted to these improvised lectures, in which he knew how to combine animated, and often eloquent, language with a judgment and art in his experiments which added to the clearness and elegance of his exposition. He exerted an actual fascination upon his auditors; and when, after having initiated them into the mysteries of science, he terminated his lecture, as he was in the habit of doing, by rising into regions far above matter, space, and time, the emotion which he experienced did not fail to communicate itself to those who listened to him, and their enthusiasm had no longer any bounds.

Before this mature and confident Faraday emerged, however, the young man was still working as a bookbinder. Faraday sent Davy a book-sized collection of notes taken from Davy's lectures and asked Davy for advice on entering the scientific field. Impressed by the young, uneducated bookbinder, Davy offered Faraday a position as his research assistant at the Royal Institute — a position that Faraday immediately accepted. Davy's letter to the Royal Institute recommending Faraday for the job revealed little hint of the greatness that awaited the unknown blacksmith's son:

> Sir Humphry Davy has the honour to inform the managers that he has found a person who is desirous to occupy the situation in the Institution lately filled by William Payne. His name is Michael Faraday. He is a youth of twenty-two years of age. As far as Sir H. Davy has been able to observe or ascertain, he appears well fitted for the situation. His habits seem good, his disposition active and cheerful, and his manner intelligent. He is willing to engage himself on the same terms as those given to Mr. Payne at the time of quitting the Institution.

Not long after Faraday's appointment as Davy's research assistant, Davy set out on a tour of Europe and asked Faraday to come along as his assistant and traveling secretary. This presented Faraday the opportunity to meet and talk with many of the great scientists of the world — contacts that he would nurture for the rest of his life. The trip, however, was not all pleasant. Just before the trip commenced, Davy asked Faraday to also serve as his valet — a service that Faraday reluctantly agreed to perform. Although Davy appears to have treated Faraday with some respect, Davy's wife, it seems, treated him with the contempt often shown by British aristocracy. Unfortunately, this was not the last time class came between Faraday and his mentor. Years later, after Faraday had established himself as a leading scientist, Davy failed to support his former assistant's nomination as a Fellow of the Royal Society. Although he won election to the Society without Davy's support, this certainly served as a reminder to Faraday that he might never be thought of as anything more than an assistant by his mentor.

After many important discoveries and publications in chemistry and electricity, Faraday finally began to come out from under the considerable shadow of Davy. Celebrated in Britain, Faraday was given many honors including an honorary degree from the University of Oxford. Faraday was appointed Fullerian Professor of Chemistry at the Royal Institution, a position he held for the remainder of his life. He gave immensely popular public lectures, especially his series of Christmas Lectures for the youth of London given at the Royal Institution. A devout Christian, Faraday was a member of the Sandemanians, a sect that broke from the Scottish Presbyterian Church and followed very strict and uncompromising rules of devotion. It was through this church that Faraday met his wife, Sarah Barnard.

Faraday's earliest scientific works were not in electricity. Davy's research interests tended towards chemistry, and as his assistant Faraday of course was expected to perform experiments in the same science. Later, however, when Davy became interested in electricity, Faraday was finally given the opportunity to delve into the field. In the opening section of Faraday's *Experimental Researches in Electricity*, the author described his purpose in conducting electrical experiments. In particular, he sought to obtain "electricity from ordinary magnetism":

> The power which electricity of tension possesses of causing an opposite electrical state in its vicinity has been expressed by the general term Induction; which, as it has been received into scientific language, may also, with propriety, be used in the same general sense to express the

power which electrical currents may possess of inducing any particular state upon matter in their immediate neighbourhood, otherwise indifferent. It is with this meaning that I purpose using it in the present paper.

Certain effects of the induction of electrical currents have already been recognised and described: as those of magnetization; Ampère's experiments of bringing a copper disc near to a flat spiral; his repetition with electro-magnets of Arago's extraordinary experiments, and perhaps a few others. Still it appeared unlikely that these could be all the effects which induction by currents could produce; especially as, upon dispensing with iron, almost the whole of them disappear, whilst yet an infinity of bodies, exhibiting definite phenomena of induction with electricity of tension, still remain to be acted upon by the induction of electricity in motion.

Further: Whether Ampère's beautiful theory were adopted, or any other, or whatever reservation were mentally made, still it appeared very extraordinary, that as every electric current was accompanied by a corresponding intensity of magnetic action at right angles to the current, good conductors of electricity, when placed within the sphere of this action, should not have any current induced through them, or some sensible effect produced equivalent in force to such a current.

These considerations, with their consequence, the hope of obtaining electricity from ordinary magnetism, have stimulated me at various times to investigate experimentally the inductive effect of electric currents. I lately arrived at positive results; and not only had my hopes fulfilled, but obtained a key which appeared to me to open out a full explanation of Arago's magnetic phenomena, and also to discover a new state, which may probably have great influence in some of the most important effects of electric currents.

These results I purpose describing, not as they were obtained, but in such a manner as to give the most concise view of the whole.

Faraday gave his law that expressed the direction of an electric current induced by the action of a magnet, or, as he called it, "magneto-electric induction":

The relation which holds between the magnetic pole, the moving wire or metal, and the direction of the current evolved, i.e. *the law* which governs the evolution of electricity by magneto-electric induction, is very simple, although rather difficult to express.

In spite of the difficulties, Faraday did express the law, providing the reader with the direction of flow for several situations in which an electric current is induced magnetically.

From a modern point of view, Michael Faraday's greatest contribution was certainly his discovery of electromagnetic induction. After years of experimentation, Faraday found that a magnet moving back and forth inside of a coil of wire produced an electric current. He continued to search for a practical way to produce electricity using his new discovery, eventually building devices that used a copper disk spinning inside the ends of a horseshoe magnet to produce a continuous electric current — the forerunner of the dynamo and the electric motor. In *Experimental Researches in Electricity*, Faraday described his first attempts to design these machines, later called Faraday disks:

Two rough trials were made with the intention of constructing *magneto-electric machines*. In one, a ring one inch and a half broad and twelve inches external diameter, cut from a thick copper plate, was mounted so as to revolve between the poles of the magnet and represent a plate similar to those formerly used ... but of interminable length; the inner and outer edges were amalgamated, and the conductors applied one to each edge, at the place of the magnetic poles. The current of electricity evolved did not appear by the galvanometer to be stronger, if so strong, as that from the circular plate....

In the other, small thick discs of copper or other metal, half an inch in diameter, were revolved rapidly near to the poles, but with the axis of rotation out of the polar axis; the electricity evolved was collected by conductors applied as before to the edge…. Currents were procured, but of strength much inferior to that produced by the circular plate.

No one, not even Faraday himself, could foresee what would come of his discovery. It would be many years before modern electric motors changed the lives of people all over the globe. Faraday, however, believed in the potential of electromagnetic induction and the devices he built. Two stories illustrate Faraday's vision. In one story, Faraday, upon demonstrating his work, was asked by either the chancellor or prime minister (the story has been told about both) "But, after all, what use is it?" to which Faraday replied, "Why sir, there is the probability that you will soon be able to tax it." The other story finds the prime minister asking Faraday about his device. "What good is it?" Faraday replied, "What good is a new-born baby?"

From a Theory of Electromagnetism to the Light Bulb

Franklin, Faraday, and other early electricians had little idea where their discoveries would lead. In the eighteenth century, electricity was an interesting philosophical (scientific) topic with very little application. The basic scientific work of Franklin and Faraday, along with that of other famous names in electricity such as Oersted, Ampère, and Coulomb, led to the inventions of Morse, Edison, and others and to the devices that much of the world now takes for granted.

Hans Christian Oersted (1777–1851) was a Danish physicist and professor at the University of Copenhagen who began his career in science performing experiments with Volta's new invention, the voltaic pile. Oersted's most important legacy was his discovery of electromagnetism, the relationship between electricity and magnetism. His revelation was perhaps the most important scientific discovery ever made in front of a classroom of students. Oersted was demonstrating an electrical experiment to his class when he noticed, quite by accident, that a wire carrying an electric current deflected the needle of a compass. Not sure what to make of his observation, Oersted repeated the experiment several months later and published his results: electric currents create magnetic fields — the birth of electromagnetism. The discovery of electromagnetism generated an immense interest among scientists, influencing Faraday, and inspired a generation to take up their own experiments in the new science.

Whereas Franklin and Faraday were brilliant experimentalists with very little mathematical ability, the French physicist André-Marie Ampère (1775–1836) was also a gifted mathematician who — inspired by the discovery of electromagnetism by Oersted — applied his skills to the science of electricity. Ampère's *Memoir on the Mathematical Theory of Electrodynamic Phenomena, Uniquely Deduced from Experience* contained the earliest mathematical treatment of the electromagnetic phenomena newly discovered by European scientists. The mathematical work of Ampère laid the groundwork for later scientists like the Scottish physicist James Clerk Maxwell (1831–1879), who developed a set of differential equations that described completely electromagnetic theory. Maxwell's mathematical description of electromagnetic fields brought together diverse discoveries and theories in magnetism, electricity, and even light into one unified theory that would have a strong influence on the development of physics in the twentieth century.

Charles-Augustin de Coulomb (1736–1806) was a French engineer, physicist, and mathematician who also made a vital contribution to the mathematical theory of electromagnetism. In particular, Coulomb discovered the mathematical law that bears his name. Coulomb's Law states that the force acting between two charged particles is inversely proportional to the distance between the two particles. This inverse square law, very similar in form to Newton's Law of Universal Gravitation, applies for both attraction and repulsion of charges.

The contributions of all the scientists mentioned in this chapter have been primarily theoretical or mathematical in nature. In our modern world, however, it is the *applications* of electricity and electromagnetism that we see all around us. From the evolution of the electric motor, to Samuel Morse's invention of the telegraph, to the electric light bulb of Thomas Edison, and the subsequent spread of electricity to households all over the world, the results of the work of Franklin, Faraday, and their contemporaries touch literally billions of lives. Perhaps no other science can make such a claim.

PRIMARY SOURCES

Davy, Humphry. "Letter recommending Faraday to the Royal Institution." *Michael Faraday: His Life and Work*, by Silvanus P. Thompson. http://books.google.com/books?id=HKf5g3qYYz8C&printsec =frontcover&source=gbs_ge_summary_r&cad=0#v=onepage&q&f=false.

Faraday, Michael. *Experimental Researches in Electricity*. Vol. 1. Project Gutenberg. http://www.gutenbe rg.org/files/14986/14986-h/14986-h.htm.

Franklin, Benjamin. *Experiments and Observations on Electricity*. London, 1751.

_____. "Experiments, observations, and facts relative to the utility of long pointed rods, for securing buildings from damage by strokes of lightning." *Memoirs of the Life and Writings of Benjamin Franklin, LL.D.* London, 1818.

Volta, Alesandro. "On the Electricity excited by the Mere Contact of Conducting Substances of Different Kinds." *Philosophical Transactions of the Royal Society*, 90 (1800): pp. 403–431. Reprinted in Baigrie, Brian. S. *Scientific Revolutions: Primary Texts in the History of Science*. Upper Saddle River, NJ: Pearson Prentice Hall, 2004.

OTHER SOURCES

Baigrie, Brian S. *Electricity and Magnetism: A Historical Perspective*. Westport, CT: Greenwood Press, 2007.

Chaplin, Joyce E. *The First Scientific American: Benjamin Franklin and the Pursuit of Genius*. New York: Basic Books, 2006.

Dray, Philip. *Stealing God's Thunder: Benjamin Franklin's Lightning Rod and the Invention of America*. New York: Random House, 2005.

Fisher, Howard J. *Faraday's Experimental Researches in Electricity: Guide to a First Reading*. Santa Fe: Green Lion Press, 2001.

Hamilton, James. *A Life of Discovery: Michael Faraday, Giant of the Scientific Revolution*. New York: Random House, 2002.

Hirshfeld, Alan. *The Electric Life of Michael Faraday*. New York: Walker, 2006.

Pancaldi, Giuliano. *Volta, Science and Culture in the Age of Enlightenment*. Princeton: Princeton University Press, 2003.

Schiffer, Michael B., Kacy L. Hollenback, and Carrie L. Bell. *Draw the Lightning Down: Benjamin Franklin and Electrical Technology in the Age of Enlightenment*. Berkeley: University of California Press, 2003.

Thomas, J. M. *Michael Faraday and the Royal Institution: The Genius of Man and Place*. Bristol, UK, Philadelphia: A. Hilger, 1991.

CHAPTER 10

Paradigm Shift: Darwin and Natural Selection

I am fully convinced that species are not immutable; but that those belonging to what are called the same genera are lineal descendants of some other and generally extinct species, in the same manner as the acknowledged varieties of any one species are the descendants of that species. Furthermore, I am convinced that Natural Selection has been the main but not exclusive means of modification. — Charles Darwin, *On the Origin of Species by Means of Natural Selection* (1859)

At the beginning of the eighteenth century, it was the general consensus of almost everyone in the Western world that the earth was 6000 years old and that the geological formations and biological species all around them were essentially the same as they had been when they were created by God. By the end of the nineteenth century, the majority of the scientific community believed that the earth was millions, or even billions, of years old and that the geological formations of the earth had slowly changed due to the inexorable natural processes of volcanic activity, erosion, and so forth. Furthermore, the theory that species had evolved, and continued to evolve, was widely accepted as a scientific principle. While this shift in human understanding of the world is generally attributed to the work of one man, Charles Darwin, many factors — and the work of many scientists — laid the groundwork for a revolution in science. Before such a revolution could occur, a fundamental shift in the science of geology was required.

Historical Geology

Traditionally, geology was concerned with collection and classification, or, in other words, the mineralogy and chemistry of the physical artifacts found in the earth. Slowly, earth scientists began to realize that the earth itself had a history, which was perhaps much different than the traditional story found in Genesis. In the late eighteenth century two competing theories emerged, with both theories claiming to explain the history of the earth.

Abraham Gottlob Werner (1750–1817) a German mining engineer, taught that water was the primary agency responsible for the formations on the surface of the earth. Rising and falling oceans, as well as erosion from rivers and runoff, all shaped the earth into its present form. Werner's followers, called Neptunists after the Roman god of the sea, were primarily catastrophists. That is, they maintained that great floods (of the order of the biblical flood survived by Noah) or other monumental upheavals unparalleled in modern times had formed the earth in a comparatively short time period. The catastrophist theory, then, fit well with the traditionalists who maintained that the earth was 6000 years old.

James Hutton (1726–1797), a Scottish naturalist and geologist, offered a different theory that credited the internal heat, or fire, of the earth as the primary formative agency. Hutton argued that this heat worked slowly and uniformly over long periods of time. Hutton's followers were known as Vulcanists after the Roman god of fire and volcanoes. More importantly, though, the theory that the formations on the earth could only have evolved slowly over a long period of time under the influence of the same natural laws currently operating on earth became known as uniformitarianism. Hutton concluded that the earth's physical characteristics could be explained only by a seemingly endless cycle of erosion, eruptions, and upheavals. He could not accept the view that the geology of the earth could be explained by a relatively recent creation followed by a relatively few catastrophic events such as the biblical flood. His poetic conclusion is one of the simplest, yet most powerful, statements from the history of science: "The result, therefore, of this physical inquiry is, that we find no vestige of a beginning,— no prospect of an end."

Although Werner had more followers — and therefore more influence — than Hutton during their lifetimes, Hutton's uniformitarianism played a vital role in the development of geology, biology, and indeed all scientific fields in the years to come.

In addition to the question of how the earth formed (and the resulting arguments concerning the age of the earth), another question that provided lively debate among natural philosophers was the nature of fossils. The meaning of the word itself indicates the confusion over what these objects really were — fossil comes from a Latin word meaning anything that was dug up. Throughout history it was usually assumed that fossils were simply rocks that resembled living organisms. How else might one explain rocks shaped like ocean creatures on a mountain peak? Or rocks with the imprint of creatures not even known to exist on earth? It was not until the seventeenth century that men like Nicholas Steno and Robert Hooke began to seriously argue that fossils were the remnants of living organisms preserved in rocks.

The true nature of fossils was finally established with the work of French naturalist Georges Cuvier. Cuvier (1769–1832) was fortunate enough to survive — and even thrive through — the French Revolution and its subsequent upheavals and executions. Cuvier used his skills in reading the fossil record and his understanding of the anatomy of living organisms to create a new science — comparative anatomy. In his work, *Essay of the Theory of the Earth* (1813), Cuvier discussed his study of remains (fossils) and his development of comparative anatomy:

I have been obliged to learn the art of deciphering and restoring these remains, of discovering and bringing together, in their primitive arrangement, the scattered and mutilated fragments of which they are composed, of reproducing, in all their original proportions and characters, the animals to which these fragments formerly belonged, and then of comparing them with those animals which still live on the surface of the earth; an art which is almost unknown, and which presupposes, what had scarcely been obtained before, an acquaintance with those laws which regulate the coexistence of the forms by which the different parts of organized beings are distinguished.

Cuvier lamented that his goal, to understand the ancient history of the globe, was hindered by the relatively primitive state of natural history:

Astronomers, no doubt, have advanced more rapidly than naturalists; and the present period, with respect to the theory of the earth, bears some resemblance to that in which some philosophers thought that the heavens were formed of polished stone, and that the moon was no larger than the Peloponnesus; but after Anaxagoras, we have had our Copernicuses, and our Keplers, who pointed out the way to Newton; and why should not natural history also have one day its Newton?

Commenting on the traditional view of fossils, Cuvier claimed the evidence was conclusive that fossils are the remains of biological organisms. In particular, Cuvier argued that fossils of seashells found far from the sea indicate an important fact about the history of the earth.

The time is past for ignorance to assert that these remains of organized bodies are mere lusus naturae,— productions generated in the womb of the earth by its own creative powers. A nice and scrupulous comparison of their forms, of their contexture, and frequently even of their composition, cannot detect the slightest difference between these shells and the shells which still inhabit the sea. They have therefore once lived in the sea, and been deposited by it; the sea consequently must have rested in the places where the deposition had taken place.

What, then, caused the tremendous upheavals required to distribute the shells in such places? Unlike Hutton, Cuvier was no uniformitarian. Cuvier argued that great events he called revolutions occurred periodically throughout Earth's history causing sudden changes to its geological features and to its inhabitants. This theory of catastrophism was used to argue against the uniformitarians who maintained that the earth's age was much older than tradition held.

In another work, *Extract from a Work on the Species of Quadrupeds of Which the Bones Have Been Found in the Interior of the Earth* (1800), Cuvier explained comparative anatomy and then made an extraordinary claim:

In the living state, all the bones are attached to each other, and form an ensemble among which all the parts are coordinated. The place that each occupies is always easy to recognize by its general form, and by the numbers and position of their articulating facets one can judge the number and direction of those that were attached to it. Now the number, direction, and shape of the bones composing each part of the body determine the movements that that part can make, and consequently the functions it can fulfill. Each part in turn in necessary relation with all the others, such that up to a certain points one can infer the ensemble from any one of them, and vice versa.

From one bone, asserted Cuvier, he can infer the structure of an entire animal! He provided a simple example:

When the teeth of an animal are such as they must be, for the animal to feed on flesh, we can be sure without further examination that the whole system of its digestive organs is adapted for this kind of food, and that its whole framework, its organs of locomotion, and even its sense organs, are made in such a way as to make it skillful in perceiving, pursuing, and seizing its prey.

In the same work, Cuvier provided concrete evidence of extinction, stating: "I have been able to restore twenty-three species, all quite certainly unknown today, and which all appear to have been destroyed, but whose existence in remote centuries is attested by their remains." He went on to describe in some detail each of these extinct species. They include mammoths, turtles, and tapirs, among others. Cuvier argued that these extinctions were caused by great revolutions, or catastrophes, in history.

The realization that fossils were indeed representative of living creatures sparked questions about extinction that scientists wrestled with for some time to come. Fossils would also eventually allow geologists to explore the history of the earth by comparing the fossil content of different strata from around the world. As naturalists gained a deeper understanding of Earth's history, some began to contemplate the history of Earth's inhabitants and to question the assumption of the immutability of species.

Evolution Before Charles Darwin

It is a common misconception that Charles Darwin was the first to propose a theory of evolution. In fact, the idea that species are not stable and unchanging predates Darwin by many years. One of the earliest proponents of the theory that modifications occur in species over a long time period was actually Charles Darwin's grandfather, Erasmus Darwin. The elder Darwin (1731–1802) was an influential physician, philosopher, and naturalist. His two-volume work, *Zoonomia, or, The Laws of Organic Life*, was an important medical text that sought to explain the basis of organic life. In one section of *Zoonomia*, Erasmus Darwin speculated about evolution, using examples and terminology that appeared in a similar vein in his famous grandson's work many decades later. This represents one of the earliest attempts to address evolutionary concepts in print. In one section, Erasmus Darwin writes about competition and survival, as well as what his grandson later called sexual selection:

> The birds, which do not carry food to their young, and do not therefore marry, are armed with spurs for the purpose of fighting for the exclusive possession of the females, as cocks and quails. It is certain that these weapons are not provided for their defence against other adversaries, because the females of these species are without this armour. The final cause of this contest amongst the males seems to be, that the strongest and most active animal should propagate the species, which should thence become improved.

He observed that various species had evidently adapted their physiology to the surroundings in a struggle to survive.

> Another great want consists in the means of procuring food, which has diversified the forms of all species of animals. Thus the nose of the swine has become hard for the purpose of

turning up the soil in search of insects and of roots. The trunk of the elephant is an elonga-
tion of the nose for the purpose of pulling down the branches of trees for his food, and for
taking up water without bending his knees. Beasts of prey have acquired strong jaws or
talons. Cattle have acquired a rough tongue and a rough palate to pull off the blades of grass,
as cows and sheep. Some birds have acquired harder beaks to crack nuts, as the parrot. Others have acquired beaks adapted to break the harder seeds, as sparrows. Others for the softer
seeds of flowers, or the buds of trees, as the finches. Other birds have acquired long beaks to
penetrate the moister soils in search of insects or roots, as woodcocks; and others broad ones
to filtrate the water of lakes, and to retain aquatic insects. All which seem to have been gradually produced during many generations by the perpetual endeavour of the creatures to supply the want of food, and to have been delivered to their posterity with constant
improvement of them for the purposes required.

Erasmus Darwin's conclusion is that it is reasonable to think that species were not
immutable, but rather changed over the long history of the earth.

> From thus meditating on the great similarity of the structure of the warm-blooded animals,
> and at the same time of the great changes they undergo both before and after their nativity;
> and by considering in how minute a portion of time many of the changes of animals above
> described have been produced; would it be too bold to imagine, that in the great length of
> time, since the earth began to exist, perhaps millions of ages before the commencement of
> the history of mankind, would it be too bold to imagine, that all warm-blooded animals
> have arisen from one living filament, which THE GREAT FIRST CAUSE endued with animality,
> with the power of acquiring new parts, attended with new propensities, directed by irrita-
> tions, sensations, volitions, and associations; and thus possessing the faculty of continuing to
> improve by its own inherent activity, and of delivering down those improvements by genera-
> tion to its posterity, world without end!

Although we find roots of evolutionary theory in the work of Erasmus Darwin, it is primarily speculative. He never attempted to provide the sort of observational proof, let
alone the mechanism, that would later be central to his grandson's theory.

Perhaps the most outspoken, and often reviled, proponent of evolution before Charles
Darwin was the French naturalist Jean-Baptiste Lamarck (1744–1829). Lamarck was a
pioneer in invertebrate zoology who argued for an evolutionary theory he called transmutation. Lamarck's evolutionary theory is summed by his two laws, which appeared in
the 1809 work *Zoological Philosophy*. The first law states:

> In every animal which has not exceeded the limit of its development, the more frequent and
> sustained use of any organ gradually strengthens this organ, develops it, makes it larger, and
> gives it a power proportional to the duration of this use; whereas, the constant lack of use of
> such an organ imperceptibly weakens it, makes it deteriorate, progressively diminishes it fac-
> ulties, and ends by making it disappear.

Lamarck used this law to argue that when an organ, or characteristic, of an animal is continuously exercised, the organ grew and strengthened. For instance, as a giraffe stretches
for the sweeter leaves higher in a tree, its neck — through continuous use — elongates. On
the other hand, if an animal possesses an organ that is not often used — a mole's eyes for
instance — that organ weakens and may eventually even disappear.

How then do these traits appear in the offspring of the giraffe with a longer neck or
a mole with weaker eyes? That is the subject of Lamarck's second law.

Everything which nature has made individuals acquire or lose through the influence of conditions to which their race has been exposed for a long time and, consequently, through the influence of the predominant use of some organ or by the influence of the constant disuse of this organ, nature preserves by reproduction in the new individuals arising from them, provided that the acquired changes are common to the two sexes or to those who have produced these new individuals.

This law, often called the inheritance of acquired traits, was the mechanism by which small changes in species led to major changes over an extended period of time.

Lamarck's theory of evolution was not well received. Powerful scientists like Georges Cuvier rejected Lamarck's evolutionary theories. Cuvier argued that there was no evidence of transmutation of species in the fossil record. He also believed that living organisms must be understood as an organic whole (or what he called the principle of correlation of parts), and that nature would not affect this balance by, for instance, changing the length of a giraffe's neck in relation to the rest of its body.

Lamarck's theory also supplied a purpose for evolutionary changes. In transmutation, organisms evolved towards more complexity, or perfection. If this were the case, critics asked, how can we explain the continued presence of simple organisms? Why have they not evolved into more complex biological species? Lamarck answered with an old theory called spontaneous generation. A continuous supply, as it were, of simple organisms appeared spontaneously from inorganic material — a process that seemed to occur whenever maggots appeared in rotting meat. Although this theory had been discredited by Francesco Redi almost two centuries earlier, Lamarck revived it to suit his own purposes.

The evolutionary theories of Erasmus Darwin, Lamarck, and others were wrought with difficulties and not generally accepted by their contemporaries. However, Charles Darwin was born into this generation in which such ideas were widely discussed. The younger Darwin was himself greatly influenced by the work of his predecessors and eventually proposed a theory of evolution that proved much more tenable to the scientific establishment.

Charles Darwin's Voyage to Discovery

Charles Darwin's (1809–1882) path to the theory that changed the world was circuitous and unlikely. Darwin's father was a well-to-do physician, and his mother was the daughter of Josiah Wedgewood, founder of the famous (and lucrative) Wedgewood pottery factory. Young Charles was a disinterested schoolboy who enjoyed spending time in nature rather than in study. As a young man, his father sent Charles to the University of Edinburgh to study medicine and thus to follow in the footsteps of Charles' brother, father, and grandfather. Charles, however, was bored with medical school, and repulsed by the sight of blood — eventually dropping out of Edinburgh. While a student at the Scottish university, however, Darwin met the zoologist Robert Grant who nurtured his interest in nature and perhaps ignited a faint ember by introducing Darwin to the evolutionary ideas of Lamarck and others.

After leaving medical school, Darwin's father arranged for his enrollment at Cambridge to begin studying for the ministry. Once again, however, Darwin's fascination with nature engulfed him and he became a disinterested theology student. However, at least one required reading while at Cambridge had a long-lasting influence on the struggling student. While reading William Paley's *Natural Theology* (1802), Darwin was struck by the logic and order of the world as described by Paley. In one famous passage, Paley explained that when one encounters a watch on the ground, the obvious implication is that it was designed by a skilled watchmaker:

> In crossing a heath, suppose I pitched my foot against a *stone*, and were asked how the stone came to be there; I might possibly answer, that, for any thing I knew to the contrary, it had lain there for ever: nor would it perhaps be very easy to show the absurdity of this answer. But suppose I had found a *watch* upon the ground, and it should be inquired how the watch happened to be in that place; I should hardly think of the answer which I had before given, that, for any thing I knew, the watch might have always been there. Yet why should not this answer serve for the watch as well as for the stone? Why is it not as admissible in the second case, as in the first? For this reason, and for no other, viz. that, when we come to inspect the watch, we perceive (what we could not discover in the stone) that its several parts are framed and put together for a purpose, e. g. that they are so formed and adjusted as to produce motion, and that motion so regulated as to point out the hour of the day; that, if the different parts had been differently shaped from what they are, of a different size from what they are, or placed after any other manner, or in any other order, than that in which they are placed, either no motion at all would have been carried on in the machine, or none which would have answered the use that is now served by it. To reckon up a few of the plainest of these parts, and of their offices, all tending to one result:— We see a cylindrical box containing a coiled elastic spring, which, by its endeavour to relax itself, turns round the box. We next observe a flexible chain (artificially wrought for the sake of flexure), communicating the action of the spring from the box to the fusee. We then find a series of wheels, the teeth of which catch in, and apply to, each other, conducting the motion from the fusee to the balance, and from the balance to the pointer; and at the same time, by the size and shape of those wheels, so regulating that motion, as to terminate in causing an index, by an equable and measured progression, to pass over a given space in a given time. We take notice that the wheels are made of brass in order to keep them from rust; the springs of steel, no other metal being so elastic; that over the face of the watch there is placed a glass, a material employed in no other part of the work, but in the room of which, if there had been any other than a transparent substance, the hour could not be seen without opening the case. This mechanism being observed (it requires indeed an examination of the instrument, and perhaps some previous knowledge of the subject, to perceive and understand it; but being once, as we have said, observed and understood), the inference, we think, is inevitable, that the watch must have had a maker: that there must have existed, at some time, and at some place or other, an artificer or artificers who formed it for the purpose which we find it actually to answer; who comprehended its construction, and designed its use.

And, Paley added, what if the watch has a peculiar characteristic: the ability to self-replicate?

> Suppose, in the next place, that the person who found the watch, should, after some time, discover that, in addition to all the properties which he had hitherto observed in it, it possessed the unexpected property of producing, in the course of its movement, another watch like itself (the thing is conceivable); that it contained within it a mechanism, a system of parts, a mould for instance, or a complex adjustment of lathes, files, and other tools, evidently and separately calculated for this purpose; let us inquire, what effect ought such a discovery to have upon his former conclusion.

The only reasonable conclusion, Paley argued, was that the watch was designed and built by a watchmaker — the created must have a creator:

> The conclusion of which the *first* examination of the watch, of its works, construction, and movement, suggested, was, that it must have had, for the cause and author of that construction, an artificer, who understood its mechanism, and designed its use. This conclusion is invincible. A *second* examination presents us with a new discovery. The watch is found, in the course of its movement, to produce another watch, similar to itself; and not only so, but we perceive in it a system or organization, separately calculated for that purpose. What effect would this discovery have, or ought it to have, upon our former inference? What, as hath already been said, but to increase, beyond measure, our admiration of the skill, which had been employed in the formation of such a machine? Or shall it, instead of this, all at once turn us round to an opposite conclusion, viz. that no art or skill whatever has been concerned in the business, although all other evidences of art and skill remain as they were, and this last and supreme piece of art be now added to the rest? Can this be maintained without absurdity? Yet this is atheism.

In the same way, Paley maintained, a natural world so beautifully designed with all of its parts fitting perfectly together must imply a designer, or creator:

> This is atheism: for every indication of contrivance, every manifestation of design, which existed in the watch, exists in the works of nature; with the difference, on the side of nature, of being greater and more, and that in a degree which exceeds all computation. I mean that the contrivances of nature surpass the contrivances of art, in the complexity, subtility, and curiosity of the mechanism; and still more, if possible, do they go beyond them in number and variety; yet, in a multitude of cases, are not less evidently mechanical, not less evidently contrivances, not less evidently accommodated to their end, or suited to their office, than are the most perfect productions of human ingenuity.

Later, Darwin reflected on Paley's arguments and reached a similar conclusion: the various components of nature were perfectly adapted to their environments. However, the mechanism by which this adaptation took place was very different for Darwin. Natural selection, not separate creation, was the cause of these endless examples of adaptation.

While still at Cambridge, perhaps the most important invitation in the history of science was extended to Darwin. Robert FitzRoy, captain of HMS *Beagle*, was set to embark on a long expedition to explore and survey lands around the world. FitzRoy sought a ship's naturalist to study the natural history of the lands he intended to survey and to provide intellectual companionship for the captain himself on the long voyage. When first approached about the opportunity, Darwin's father denied him permission, thinking the voyage a waste of time and just another lark by his immature son. However, a letter from his brother-in-law, Josiah Wedgewood II, convinced the elder Darwin that the voyage was a rare opportunity for his son as a budding naturalist. Thus began a journey that eventually led Charles Darwin to change the very foundation of science.

When the *Beagle* set sail in 1831, Charles Darwin was a young man with no career and seemingly little direction. When the ship finally returned to England five years later, Darwin was already beginning to build a reputation as an accomplished naturalist. For those five years, as the *Beagle* literally sailed around the world, Darwin explored the lands (especially South America), observing, collecting, and thinking about what he observed. He was also fortunate that he had a copy of the first volume of Charles Lyell's *Principles of Geology,* a work that inspired him to begin thinking about geological time and the principles of uniformitarianism. In his highly influential treatise, Lyell defined geology much differently than tradition dictated. Historically, geology was the study of the static features of the earth: mineralogy, chemistry, and other physical features of the planet. Lyell, building on the uniformitarian principles of James Hutton, helped to redefine geology as the study of the history of the earth:

> Geology is the science which investigates the successive changes that have taken place in the organic and inorganic kingdoms of nature; it inquires into the causes of these changes, and the influence which they have exerted in modifying the surface and external structure of our planet.

Lyell's work, along with Darwin's own observations, convinced the young naturalist that land masses were built up over extremely long periods of time. These geological ideas also began to open up other avenues of thought for Darwin — if the geological features of the earth slowly evolved over long periods of time, why not the biological systems?

Throughout the long voyage, it became common for Captain FitzRoy to leave Darwin on shore as the *Beagle* surveyed and mapped the coastline. Darwin ventured deep into the interior of the land, accompanied by selected shipmates and local guides. In South America, for instance, Darwin spent weeks exploring the interior with gauchos, collecting a wide range of specimens to be crated and shipped back to England. At this point, Darwin was not yet an expert in botany, zoology, or any other branch of natural history, so these specimens were intended for study and classification by experts at home.

Although the *Beagle* was only there a short time, a small archipelago off the western coast of Ecuador, known as the Galapagos Islands, proved crucial to Darwin's eventual understanding of several principles central to his theories of evolution. Although it is only one of the many important observations made by Darwin while on the islands, his collection of finches provides perhaps the most famous discovery of the entire voyage. In particular, the beaks of the finches on the various islands varied depending on the food supply available. Interestingly, Darwin did not immediately recognize the importance of the Galapagos finches. In fact, he did not even realize they were all finches until his specimens were examined by an ornithologist in London, John Gould. Once Gould made it clear to Darwin that these birds were all different varieties of finches, he began thinking about how new species appeared.

In his published account of the expedition, *Voyage of the Beagle* (1839), Darwin related his observations concerning the birds of the Galapagos. He began with a general discussion of what he observed, including several species that did not seem to strike Darwin as particularly significant.

> Of land-birds I obtained twenty-six kinds, all peculiar to the group and found nowhere else, with the exception of one lark-like finch from North America (Dolichonyx oryzivorus), which ranges on that continent as far north as 54 degs., and generally frequents marshes. The other twenty-five birds consist, firstly, of a hawk, curiously intermediate in structure between a buzzard and the American group of carrion-feeding Polybori; and with these latter birds it agrees most closely in every habit and even tone of voice. Secondly, there are two owls, representing the short-eared and white barn-owls of Europe. Thirdly, a wren, three tyrant-flycatchers (two of them species of Pyrocephalus, one or both of which would be ranked by

some ornithologists as only varieties), and a dove — all analogous to, but distinct from, American species. Fourthly, a swallow, which though differing from the Progne purpurea of both Americas, only in being rather duller colored, smaller, and slenderer, is considered by Mr. Gould as specifically distinct. Fifthly, there are three species of mocking thrush — a form highly characteristic of America.

Darwin proceeded to describe what he later came to realize was the large group of finches on the various islands. In particular, Darwin was struck by the apparent adaptation of the beaks of the finches to the local food supply.

> The remaining land-birds form a most singular group of finches, related to each other in the structure of their beaks, short tails, form of body and plumage: there are thirteen species, which Mr. Gould has divided into four sub-groups. All these species are peculiar to this archipelago; and so is the whole group, with the exception of one species of the sub-group Cactornis, lately brought from Bow Island, in the Low Archipelago. Of Cactornis, the two species may be often seen climbing about the flowers of the great cactus-trees; but all the other species of this group of finches, mingled together in flocks, feed on the dry and sterile ground of the lower districts. The males of all, or certainly of the greater number, are jet black; and the females (with perhaps one or two exceptions) are brown. The most curious fact is the perfect gradation in the size of the beaks in the different species of Geospiza, from one as large as that of a hawfinch to that of a chaffinch, and (if Mr. Gould is right in including his sub-group, Certhidea, in the main group) even to that of a warbler. The largest beak in the genus Geospiza is shown in Fig. 1, and the smallest in Fig. 3; but instead of there being only one intermediate species, with a beak of the size shown in Fig. 2, there are no less than six species with insensibly graduated beaks. The beak of the sub-group Certhidea, is shown in Fig. 4. The beak of Cactornis is somewhat like that of a starling, and that of the fourth sub-group, Camarhynchus, is slightly parrot-shaped. Seeing this gradation and diversity of structure in one small, intimately related group of birds, one might really fancy that from an original paucity of birds in this archipelago, one species had been taken and modified for different ends. In a like manner it might be fancied that a bird originally a buzzard, had been induced here to undertake the office of the carrion-feeding Polybori of the American continent.

In the last lines of this entry, we see Darwin making a tentative proposal that such characteristics might be explained by the modification of one species, a claim that he made much more forcefully in years to come.

Upon return from his epic voyage, Darwin settled into the life of a gentleman naturalist, working on many projects considered mainstream biology, from pigeons to barnacles. In *The Structure and Distribution of Coral Reefs* (1842), for instance, Darwin used some of the numerous observations he made while aboard the *Beagle* to propose theories concerning the formation of coral reefs and atolls. Slowly and steadily, Darwin began building a reputation as a first rate naturalist.

Meanwhile, the young man began contemplating marriage, even though he had not yet found a woman who interested him. Sounding very much like the careful, methodical scientist he was becoming, Darwin jotted down some of the reasons both for and against marriage. Included among the reasons not to marry were:

> If marry — means limited. Feel duty to work for money. London life, nothing but Society, no country, no tours, no large Zoolog: collect., no books.
> Freedom to go where one liked. Choice of Society *and little of it*. Conversation of clever

men at clubs. Not forced to visit relatives, and to bend in every trifle — to have the expense and anxiety of children — perhaps quarrelling. *Loss of time* — cannot read in the evenings — fatness and idleness — anxiety and responsibility — less money for books etc.

On the other hand, Darwin found some compelling reasons to marry:

Children — (if it please God) — constant companion, (friend in old age) who will feel interested in one, object to be beloved and played with — better than a dog anyhow — Home, and someone to take care of house — Charms of music and female chit-chat. These things good for one's health.

My God, it is intolerable to think of spending one's whole life, like a neuter bee, working, working and nothing after all. No, no won't do. Imagine living all one's day solitarily in smoky dirty London House. — Only picture to yourself a nice soft wife on a sofa with good fire, and books and music perhaps.... Marry — Marry — Marry Q.E.D.

Three years after the end of his great adventure aboard the *Beagle*, Darwin took his own advice and married his cousin, Emma Wedgewood. Between her dowry and income from his father, Darwin never worried about supporting himself or his bride and was able to spend his time pursuing science.

Origin of Species

Charles Darwin's epic adventure on the *Beagle* ended in 1836; yet, his epic-making book, *Origin of Species* (whose full title was *On the Origin of Species by Means of Natural Selection, or the Preservation of Favoured Races in the Struggle for Life*), was not published until 1859, over two decades later. Why the interminable delay? The reasons are many, and complex — even debatable. Part of the reason was that Darwin needed to establish himself as a legitimate scientist before he proposed such a revolutionary theory: thus his work on barnacles, coral reefs, and other subjects in conventional biology. Darwin knew, of course, that what he was to propose would be controversial, probably even inflammatory. So, in addition to establishing his reputation, Darwin spent these years collecting and organizing the data and the arguments that supported his theories. Darwin related these years of work to his readers, in the introduction to *Origin of Species*:

When on board H.M.S. "Beagle," as naturalist, I was much struck with certain facts in the distribution of the inhabitants of South America, and in the geological relations of the present to the past inhabitants of that continent. These facts seemed to me to throw some light on the origin of species — that mystery of mysteries, as it has been called by one of our greatest philosophers. On my return home, it occurred to me, in 1837, that something might perhaps be made out on this question by patiently accumulating and reflecting on all sorts of facts which could possibly have any bearing on it. After five years' work I allowed myself to speculate on the subject, and drew up some short notes; these I enlarged in 1844 into a sketch of the conclusions, which then seemed to me probable: from that period to the present day I have steadily pursued the same object. I hope that I may be excused for entering on these personal details, as I give them to show that I have not been hasty in coming to a decision.

Finally, Darwin hesitated (granted it was a long hesitation) because he was not anxious to dive into such controversy, especially in light of the challenge to orthodox Christian views (his own wife was a devout Christian) he knew his theories would bring.

Add to the mix difficulties ranging from his own bouts with poor health and the death of a child, and we see that Darwin had many distractions drawing his attention away from evolution.

With these distractions ongoing, what finally prompted Darwin to begin seriously composing his ideas and theories into a publishable form? As it turns out, Darwin was not the only naturalist gathering evidence for evolutionary theories, as he relates in *Origin of Species*:

> My work is now nearly finished; but as it will take me two or three more years to complete it, and as my health is far from strong, I have been urged to publish this Abstract. I have more especially been induced to do this, as Mr. Wallace, who is now studying the natural history of the Malay archipelago, has arrived at almost exactly the same general conclusions that I have on the origin of species. Last year he sent to me a memoir on this subject, with a request that I would forward it to Sir Charles Lyell, who sent it to the Linnean Society, and it is published in the third volume of the Journal of that Society. Sir C. Lyell and Dr. Hooker, who both knew of my work — the latter having read my sketch of 1844 — honoured me by thinking it advisable to publish, with Mr. Wallace's excellent memoir, some brief extracts from my manuscripts.

Alfred Russell Wallace (1823–1913) was a British naturalist who explored and collected in Amazonia before moving on to Southeast Asia, specifically the Malay Archipelago. Wallace spent many years making observations and eventually coming to conclusions remarkably similar to those drawn by Darwin. When an essay outlining these theories reached Darwin, with Wallace's request to share the essay with Charles Lyell, Darwin was prompted to rather hurriedly compose his own essay and Lyell ensured that the work of both men was published together.

In is essay, Wallace outlined arguments very similar to those Darwin had been compiling for many years. He wrote of the struggle for existence, a theory that would later be called survival of the fittest:

> The life of wild animals is a struggle for existence. The full exertion of all their faculties and all their energies is required to preserve their own existence and provide for that of their infant offspring. The possibility of procuring food during the least favourable seasons, and of escaping the attacks of their most dangerous enemies, are the primary conditions which determine the existence both of individuals and of entire species. These conditions will also determine the population of a species; and by a careful consideration of all the circumstances we may be enabled to comprehend, and in some degree to explain, what at first sight appears so inexplicable — the excessive abundance of some species, while others closely allied to them are very rare.
>
> Now it is clear that what takes place among the individuals of a species must also occur among the several allied species of a group, — viz. that those which are best adapted to obtain a regular supply of food, and to defend themselves against the attacks of their enemies and the vicissitudes of the seasons, must necessarily obtain and preserve a superiority in population; while those species which from some defect of power or organization are the least capable of counteracting the vicissitudes of food, supply, &c., must diminish in numbers, and, in extreme cases, become altogether extinct. Between these extremes the species will present various degrees of capacity for ensuring the means of preserving life; and it is thus we account for the abundance or rarity of species.

Wallace also argued for a similar mechanism for evolution as had Darwin, one that opposed Lamarck's theory of acquired traits:

The powerful retractile talons of the falcon- and the cat-tribes have not been produced or increased by the volition of those animals; but among the different varieties which occurred in the earlier and less highly organized forms of these groups, *those always survived longest which had the greatest facilities for seizing their prey.* Neither did the giraffe acquire its long neck by desiring to reach the foliage of the more lofty shrubs, and constantly stretching its neck for the purpose, but because any varieties which occurred among its antitypes with a longer neck than usual *at once secured a fresh range of pasture over the same ground as their shorter-necked companions, and on the first scarcity of food were thereby enabled to outlive them.*

Wallace's concluding paragraph in this short communication summarized his theory:

We believe we have now shown that there is a tendency in nature to the continued progression of certain classes of *varieties* further and further from the original type — a progression to which there appears no reason to assign any definite limits — and that the same principle which produces this result in a state of nature will also explain why domestic varieties have a tendency to revert to the original type. This progression, by minute steps, in various directions, but always checked and balanced by the necessary conditions, subject to which alone existence can be preserved, may, it is believed, be followed out so as to agree with all the phenomena presented by organized beings, their extinction and succession in past ages, and all the extraordinary modifications of form, instinct, and habits which they exhibit.

Even as Darwin hurriedly compiled a short communication on his own theories to be published along with Wallace's letter, he also began to work in earnest to compose a larger work, one that would shortly be published as *Origin of Species* in 1859.

Darwin laid out his theory of descent clearly in *Origin of Species*:

In considering the Origin of Species, it is quite conceivable that a naturalist, reflecting on the mutual affinities of organic beings, on their embryological relations, their geographical distribution, geological succession, and other such facts, might come to the conclusion that each species had not been independently created, but had descended, like varieties, from other species. Nevertheless, such a conclusion, even if well founded, would be unsatisfactory, until it could be shown how the innumerable species inhabiting this world have been modified, so as to acquire that perfection of structure and coadaptation which most justly excites our admiration.

Note that Darwin was careful not to use the term *evolution*, but rather wrote in terms of descent, modification, and adaptation. In the nineteenth century, the term evolution was vague and had different meanings. For clarity, Darwin chose the phrase "descent with modification" throughout his work to describe the theory.

Darwin continued to explain his primary conclusion, one that he supported throughout *Origin of Species* with data and observations gathered over a lifetime of work.

I am fully convinced that species are not immutable; but that those belonging to what are called the same genera are lineal descendants of some other and generally extinct species, in the same manner as the acknowledged varieties of any one species are the descendants of that species. Furthermore, I am convinced that Natural Selection has been the main but not exclusive means of modification.

Natural selection, of course, is universally associated with Darwin. From where, though, did the initial idea come? Darwin shared the seeds of a scientific theory with his readers:

This is the doctrine of Malthus, applied to the whole animal and vegetable kingdoms. As many more individuals of each species are born than can possibly survive; and as, consequently, there is a frequently recurring struggle for existence, it follows that any being, if it vary however slightly in any manner profitable to itself, under the complex and sometimes varying conditions of life, will have a better chance of surviving, and thus be NATURALLY SELECTED. From the strong principle of inheritance, any selected variety will tend to propagate its new and modified form.

The Rev. Thomas Malthus (1766–1834) was a British economist who wrote a very influential — and controversial — book called *Essay on the Principle of Population.* In this work, Malthus theorized that human population grew exponentially while food production increased linearly. This meant that, inevitably, there would be severe shortages of food so that famine (or disease or social maladies resulting from the food shortage) would act as a check on population growth:

> I say, that the power of population is indefinitely greater than the power in the earth to produce subsistence for man.
>
> Population, when unchecked, increases in a geometrical ratio. Subsistence increases only in an arithmetical ratio. A slight acquaintance with numbers will shew the immensity of the first power in comparison of the second.
>
> By that law of our nature which makes food necessary to the life of man, the effects of these two unequal powers must be kept equal.
>
> This implies a strong and constantly operating check on population from the difficulty of subsistence. This difficulty must fall somewhere and must necessarily be severely felt by a large portion of mankind.

Contemplating the consequences of Malthus' theory on biological populations other than humans, Darwin came to the conclusion that nature was all about a struggle for survival. He based his theory of natural selection on this never-ending struggle. The term most associated with the winners of this struggle is survival of the fittest, a term actually coined by Herbert Spencer some seven years after *Origin of Species* was published.

Looking back at evolutionary theories such as those proposed by Lamarck, one of the primary reasons for their rejection was the lack of a suitable mechanism for the evolutionary process. Darwin's mechanism was both simple and elegant:

> Owing to this struggle for life, any variation, however slight and from whatever cause proceeding, if it be in any degree profitable to an individual of any species, in its infinitely complex relations to other organic beings and to external nature, will tend to the preservation of that individual, and will generally be inherited by its offspring. The offspring, also, will thus have a better chance of surviving, for, of the many individuals of any species which are periodically born, but a small number can survive. I have called this principle, by which each slight variation, if useful, is preserved, by the term of Natural Selection, in order to mark its relation to man's power of selection.

Later, Darwin wrote:

> This preservation of favourable variations and the rejection of injurious variations, I call Natural Selection.
>
> ... I can see no reason to doubt that an accidental deviation in the size and form of the body, or in the curvature and length of the proboscis, etc., far too slight to be appreciated by us, might profit a bee or other insect, so that an individual so characterised would be able to obtain its food more quickly, and so have a better chance of living and leaving descendants. Its descendants would probably inherit a tendency to a similar slight deviation of structure.

In one line of the passage above, Darwin alluded to "man's power of selection." He was referring here to domestication of plants and animals by man, and Darwin used his vast knowledge of domestication to create a powerful analogy to natural selection.

Darwin employed analogies with the domestication of animals very familiar to typical Englishmen to show how man has selected for certain characteristics useful to him: "One of the most remarkable features in our domesticated races is that we see in them adaptation, not indeed to the animal's or plant's own good, but to man's use or fancy."

He discussed the differences in horses (from thoroughbreds for racing to plow horses bred for strength and stamina), dogs, and especially an animal of interest to many a British breeder, pigeons.

> Great as the differences are between the breeds of pigeons, I am fully convinced that the common opinion of naturalists is correct, namely, that all have descended from the rock-pigeon (Columba livia), including under this term several geographical races or sub-species, which differ from each other in the most trifling respects.
>
> I have never met a pigeon, or poultry, or duck, or rabbit fancier, who was not fully convinced that each main breed was descended from a distinct species.

Darwin then drew the analogy between domestication (selection influenced by humans, not nature) and natural selection:

> We have seen that man by selection can certainly produce great results, and can adapt organic beings to his own uses, through the accumulation of slight but useful variations, given to him by the hand of Nature. But Natural Selection, as we shall hereafter see, is a power incessantly ready for action, and is as immeasurably superior to man's feeble efforts, as the works of Nature are to those of Art.
>
> Can it, then, be thought improbable, seeing that variations useful to man have undoubtedly occurred, that other variations useful in some way to each being in the great and complex battle of life, should sometimes occur in the course of thousands of generations? If such do occur, can we doubt (remembering that many more individuals are born than can possibly survive) that individuals having any advantage, however slight, over others, would have the best chance of surviving and of procreating their kind?

Having already compared man's "feeble efforts" to those of nature, Darwin concluded that nature has the power, and more importantly the time, to cause even greater change.

> Slow though the process of selection may be, if feeble man can do much by his powers of artificial selection, I can see no limit to the amount of change, to the beauty and infinite complexity of the coadaptations between all organic beings, one with another and with their physical conditions of life, which may be effected in the long course of time by nature's power of selection.
>
> If during the long course of ages and under varying conditions of life, organic beings vary at all in the several parts of their organisation, and I think this cannot be disputed; if there be, owing to the high geometrical powers of increase of each species, at some age, season, or year, a severe struggle for life, and this certainly cannot be disputed; then, considering the infinite complexity of the relations of all organic beings to each other and to their conditions of existence, causing an infinite diversity in structure, constitution, and habits, to be advantageous to them, I think it would be a most extraordinary fact if no variation ever had occurred useful to each being's own welfare, in the same way as so many variations have occurred useful to man.

Descent with modification, powered by natural selection, was the primary theme of *Origin of Species*. It was not, however, the only subject broached by Darwin. Taking advantage of more data and observations accumulated over many decades, Darwin proposed his theory of sexual selection. Whereas natural selection spoke to the survival of a species, sexual selection, as Darwin explained, was much more a function of ensuring offspring from individuals in the species.

> And this leads me to say a few words on what I call Sexual Selection. This depends, not on a struggle for existence, but on a struggle between the males for possession of the females; the result is not death to the unsuccessful competitor, but few or no offspring. Sexual selection is, therefore, less rigorous than natural selection. Generally, the most vigorous males, those which are best fitted for their places in nature, will leave most progeny. But in many cases, victory will depend not on general vigour, but on having special weapons, confined to the male sex. A hornless stag or spurless cock would have a poor chance of leaving offspring.

Darwin's publication of *Origin of Species* was met with mixed reactions, but surprise was certainly not among those reactions. As we have seen, evolution was not an original idea for Darwin. He followed Lamarck, and even his own grandfather, Erasmus Darwin, among others in professing that species were not immutable. However, for the first time Darwin had presented a workable mechanism for evolution — natural selection.

Responses to *Origin of Species* ran the gamut from immediate acceptance to outright rejection, with many leading scientists falling somewhere in the middle. Thomas Huxley became known as "Darwin's Bulldog" for his tenacious defense of the theory. Influential scientists such as the geologist Adam Sedgwick and naturalist Louis Agassiz were among the outspoken critics of Darwin's work. A large number of scientists, however, such as Charles Lyell were cautious in their analysis of Darwin and remained open to, if not fully convinced of, the validity of his theory.

In one famous exchange, Bishop Samuel Wilberforce engaged Huxley and Joseph Hooker in an open debate over the implications of Darwin's work. In this impromptu debate, held after a lecture given before the British Association for the Advancement of Science, it was reported that Wilberforce turned to Huxley and asked whether he was descended from apes through his grandfather or grandmother. Huxley's response was (reportedly) that he would not be ashamed to have a monkey for an ancestor, but he would be ashamed to be connected with a man who used great gifts to obscure the truth. Whether the debate proceeded exactly as reported, it certainly brought to the forefront the argument from Darwin's supporters that theologians had no business trying to determine the direction of scientific inquiry.

In the years after publication of *Origin of Species*, and in the ensuing 150 years, Darwin has often been vilified for removing humankind from its special place in God's creation. Interestingly, Darwin did not include humans in *Origin of Species*. In fact, the only mention came in the rather cryptic line, "Light will be thrown on the origin of man and his history." Darwin himself did not shed further light on the origin of man for another dozen years. In the meantime, other works claiming the place of humans in evolution appeared.

In 1871, Darwin published his second major work on evolutionary theory, *The Descent of Man, and Selection in Relation to Sex*. In the introduction, Darwin related why he delayed another twelve years the publication of this work.

The nature of the following work will be best understood by a brief account of how it came to be written. During many years I collected notes on the origin or descent of man, without any intention of publishing on the subject, but rather with the determination not to publish, as I thought that I should thus only add to the prejudices against my views. It seemed to me sufficient to indicate, in the first edition of my "Origin of Species," that by this work "light would be thrown on the origin of man and his history;" and this implies that man must be included with other organic beings in any general conclusion respecting his manner of appearance on this earth.

In consequence of the views now adopted by most naturalists, and which will ultimately, as in every other case, be followed by others who are not scientific, I have been led to put together my notes, so as to see how far the general conclusions arrived at in my former works were applicable to man. This seemed all the more desirable, as I had never deliberately applied these views to a species taken singly. When we confine our attention to any one form, we are deprived of the weighty arguments derived from the nature of the affinities which connect together whole groups of organisms — their geographical distribution in past and present times, and their geological succession. The homological structure, embryological development, and rudimentary organs of a species remain to be considered, whether it be man or any other animal, to which our attention may be directed; but these great classes of facts afford, as it appears to me, ample and conclusive evidence in favour of the principle of gradual evolution. The strong support derived from the other arguments should, however, always be kept before the mind.

The conclusion that man is the co–descendant with other species of some ancient, lower, and extinct form, is not in any degree new. Lamarck long ago came to this conclusion, which has lately been maintained by several eminent naturalists and philosophers; for instance, by Wallace, Huxley, Lyell, Vogt, Lubbock, Buchner, Rolle, etc.

As the title implies, the subject of Darwin's 1871 work was two-pronged: man's place in evolution and the role of sexual selection applied to various races of humans.

During many years it has seemed to me highly probable that sexual selection has played an important part in differentiating the races of man; but in my "Origin of Species" (first edition, page 199) I contented myself by merely alluding to this belief. When I came to apply this view to man, I found it indispensable to treat the whole subject in full detail.... Consequently the second part of the present work, treating of sexual selection, has extended to an inordinate length, compared with the first part; but this could not be avoided.

In *Descent of Man*, Darwin asked (and answered) the basic question of

whether man tends to increase at so rapid a rate, as to lead to occasional severe struggles for existence; and consequently to beneficial variations, whether in body or mind, being preserved, and injurious ones eliminated. Do the races or species of men, whichever term may be applied, encroach on and replace one another, so that some finally become extinct? We shall see that all these questions, as indeed is obvious in respect to most of them, must be answered in the affirmative, in the same manner as with the lower animals.

And Darwin concluded that

the main conclusion arrived at in this work, namely, that man is descended from some lowly organised form, will, I regret to think, be highly distasteful to many.

Man may be excused for feeling some pride at having risen, though not through his own exertions, to the very summit of the organic scale; and the fact of his having thus risen, instead of having been aboriginally placed there, may give him hope for a still higher destiny in the distant future. But we are not here concerned with hopes or fears, only with the truth as far as our reason permits us to discover it; and I have given the evidence to the best of my ability. We must, however, acknowledge, as it seems to me, that man with all his noble qualities, with sympathy which feels for the most debased, with benevolence which extends not only to other men but to the humblest living creature, with his god-like intellect which has penetrated into the movements and constitution of the solar system — with all these exalted powers — Man still bears in his bodily frame the indelible stamp of his lowly origin.

Darwin's Impact

As we have seen, Charles Darwin did not have the first word in evolution. As it turns out, he had far from the last word. The twentieth century fusion of Darwinian evolution with a modern understanding of genetics led to a fundamental shift in the way scientists understand life itself. Armed with these tools, along with discoveries such as the role of DNA in replicating life, has led to scientific advances Darwin could never have imagined.

The now ubiquitous term paradigm shift was introduced by Thomas Kuhn in his book *The Structure of Scientific Revolutions* (1962). Kuhn restricted his use of the term to fundamental changes in a scientific discipline which require wholesale changes in the fundamental tenets of the discipline. Since that time, however, various authors have broadened the scope of the term paradigm shift to include fundamental changes in how those outside of science understand the world. Many episodes in the history of science might qualify as paradigm shifts (Copernicanism, Newtonianism, etc.), but perhaps none have so fundamentally affected humankind's perception of the world — and themselves — as Darwinianism. The implications of Darwinian evolution for biology, philosophy, theology, sociology, anthropology, and a host of other -ologies has firmly planted Darwin in the psyche of modern humans and forever changed the way we think about our world.

PRIMARY SOURCES

Cuvier, Georges. *Essay of the Theory of the Earth*, translated by Robert Jameson. New York: Kirk & Mercein, 1818.

_____. *Extract from a Work on the Species of Quadrupeds of Which the Bones Have Been Found in the Interior of the Earth. Georges Cuvier, Fossil Bones, and Geological Catastrophes*, by Martin J. S. Rudwick. Chicago: University of Chicago Press, 1997.

Darwin, Charles. *Origin of Species*. London, 1859.

_____. *Voyage of the Beagle*. London, 1845.

Darwin, Erasmus. *Zoonomia, or, The Laws of Organic Life* Vol. 1. Project Gutenberg. http://www.gutenberg.org/files/15707/15707-h/15707-h.htm.

Hutton, James. *The Theory of the Earth*. Vol. 1. Project Gutenberg. http://www.gutenberg.org/files/12861/12861-h/12861-h.htm.

Lamarck, J.B. *Zoological Philosophy*, translated by Ian Johnston, Malaspina University-College, Nanaimo, BC, Canada. http://records.viu.ca/~johnstoi/lamarck/tofc.htm.

Lyell, Charles. *Principles of Geology*. London, 1830.

Malthus, Thomas. *Essay on Population*. Project Gutenberg. http://www.gutenberg.org/files/4239/4239-h/4239-h.htm.

Paley, William. *Natural Theology; or, Evidences of the Existence and Attributes of the Deity*. http://darwin-online.org.uk.

Wallace, Alfred Russell. "On the Tendency of Varieties to depart indefinitely from the Original Type." *Journal of the Proceedings of the Linnean Society* vol. 3 (1859).

OTHER SOURCES

Adams, Frank Dawson. *The Birth and Development of the Geological Sciences*. New York: Dover, 1938.

Amigoni, David, and Jeff Wallace. *Charles Darwin's the Origin of Species: New Interdisciplinary Essays*. Manchester: Manchester University Press, 1995.

Appleman, Philip, ed. *Darwin. Norton Critical Edition*. New York: W.W. Norton, 1970.

Aydon, Cyril. *Charles Darwin: The Naturalist Who Started a Scientific Revolution*. New York: Carroll & Graf, 2002.

Bowlby, John. *Charles Darwin: A New Life*. New York: W.W. Norton, 1991.

Clark, Ronald William. *The Survival of Charles Darwin: A Biography of a Man and an Idea*. New York: Random House, 1984.

De Beer, Gavin. *Charles Darwin; Evolution by Natural Selection*. London and New York: T. Nelson, 1963.

Desmond, Adrian J., James R. Moore, and E. J. Browne. *Charles Darwin*. Oxford and New York: Oxford University Press, 2007.

Francis, Keith. *Charles Darwin and The Origin of Species*. Westport, CT: Greenwood Press, 2007.

Herbert, Sandra. *Charles Darwin, Geologist*. Ithaca: Cornell University Press, 2005.

Keynes, R. D. *Fossils, Finches and Fuegians: Charles Darwin's Adventures and Discoveries on the Beagle, 1832–1836*. London: HarperCollins, 2002.

Laudan, Rachel. *From Mineralogy to Geology: The Foundations of a Science, 1650–1830*. Chicago: University of Chicago Press, 1987.

Mayr, Ernst. *One Long Argument: Charles Darwin and the Genesis of Modern Evolutionary Thought*. Cambridge: Harvard University Press, 1991.

CHAPTER 11

Laplace to Galton: Uncertainty in the Physical and Social Sciences

Strictly speaking it may even be said that nearly all our knowledge is problematical; and in the small number of things which we are able to know with certainty, even in the mathematical sciences themselves, the principal means for ascertaining truth — induction and analogy — are based on probabilities. — Pierre Laplace, *A Philosophical Essay on Probabilities* (1825)

The problems of family likeness do not admit of being properly expressed except in the technical language of the laws of chance, and that it is impossible to discuss them adequately except through the medium of mathematics. — Francis Galton, "Kinship and Correlation" (1890)

Galileo, Newton, Kepler, and their contemporaries sought to explain the world around them in precise and exact terms. Most believed that their duty as natural philosophers was to discover the laws of nature as they were authored by the creator. Because God was perfection in every aspect, these laws, of course, must be universal and constant (except in the rare instances where miraculous intervention occurred). Even the scientists of the Enlightenment who denied the existence of God — or at least the omnipresence of God in the natural world — sought after the same sort of consistency in nature. Predictability, predicated on exact mathematical descriptions of phenomena, was the goal of science and scientists. Because of this mindset, it was very difficult for these men and their successors to accept the possibility that nature acted randomly and that predictability must give way to probability when analyzing natural phenomena. If the scientific revolution was the hallmark of seventeenth and eighteenth century science, the probabilistic revolution — or, as it has been called, the death of certainty — might be called the hallmark of science in the nineteenth and early twentieth century.

So, the rise of probabilistic thinking may be counted as one of the most important and far-reaching accomplishments in science during the nineteenth century. In addition to the monumental advances in the mathematics of probability, the changes in the way the natural world was understood had a lasting influence upon science and society as a whole. It may be said that at the beginning of the century, science was the search for universal and absolute truths, while by the end of the century, the scientific community was coming to terms with the idea that much of science could only be known to varying degrees of certainty as measured by probability.

Although probability and statistics are branches of mathematics, their application to the sciences — and perhaps more importantly their role in changing the very perception of the goals of science — makes them critical for understanding modern science. Basic ideas in probability are as intuitive as arithmetic. If it is easy for us to imagine prehistoric man counting, it is just as easy to imagine him employing intuitive probability. A hunter, for instance, must decide what prey to chase. A small deer or rabbit presents little danger, but also may not provide enough meat for his family or tribe. Larger and more dangerous prey, on the other hand, might provide food for days or even weeks. Is the danger worth the risk? What is the probability of success? What is the probability of injury or death for the hunter? Obviously, the primitive hunter did not think in such sophisticated terms; yet, he still weighed these probabilities from his experiences before the hunt commenced. It would be the modern era, however, before probability as intuition became quantified and understood by mathematicians and natural philosophers.

Probability Theory Is Born

The eighteenth century was a time of great expectation for science. Using the incredible discoveries of Isaac Newton as a starting point, scientists (still called natural philosophers until well into the nineteenth century) began to discover great regularities in all parts of the natural world. The success found in utilizing Newtonian science to describe phenomena in the physical sciences led those working in fields dealing with living things, with minerals, with the earth, and even with social phenomena, to believe that all such fields might be quantified along the same lines as the physical sciences.

During this flurry of quantification, the field of probability began to take shape. The roots of mathematical probability are usually traced to a series of letters exchanged between the French mathematicians Pierre de Fermat (1601–1665) and Blaise Pascal (1623–1662). The topic of the letters was an activity almost as old as civilization itself— gambling. In particular, the following question was posed to Pascal by a friend interested in gambling and mathematics. Suppose two friends play a game of chance (rolling dice, for instance). The winner of the game — and therefore of the wager made by the players — is the first to win three times. Suppose further that the game was interrupted after it began but before one of the players had won three times. How should the wager be divided? If, for instance, player one is ahead two games to one, what portion of the wager should player one receive in recognition of his greater chance of winning three games?

Blaise Pascal thought he had the answer. He composed his thoughts in a letter to Pierre de Fermat, who soon posted a return letter with his own ideas concerning the solution to the problem. Over the next few months the two great mathematicians exchanged a series of letters that formed the basis for the mathematical theory of probability. Who were these men who unknowingly began one of the great revolutions in science?

Blaise Pascal, was a French mathematician, physicist, and philosopher who is perhaps best remembered for Pascal's Triangle, used by students even today to find the coefficients of the expansion of any binomial. Following is the first six rows of Pascal's Triangle.

```
                      1
              1               1
         1           2            1
      1          3          3         1
    1         4         6         4        1
  1        5        10        10        5        1
```

Each row begins and ends with one. The other numbers are simply the sum of the two numbers directly above. For instance, the number 4 in row five is 1 + 3. The next number, 6, is just 3 + 3. With that knowledge, rows may be constructed indefinitely. Although this simple arrangement of numbers was known in cultures from China to Persia and predated Pascal by centuries, his work influenced its spread throughout Europe.

As a young man, Pascal also invented one of the earliest mechanical calculators to help his father (a tax collector) with his calculations. Known as the pascaline, the calculator was a technical success; unfortunately, it was a commercial failure.

Later in life, Pascal gave up his scientific and mathematical interests to pursue theological questions. In one very unique process, Pascal argued that a belief in God was not only theologically sound, but also mathematically sound. Pascal essentially calculated the mathematical expectation of a belief in God. Simply put, Pascal claimed one could choose to either believe in God or not believe in God. Of course, God either does exist or does not. Now, if one believes in God yet God does not exist, the outcome is neutral. The person led a good life but receives nothing for it. If one does not believe in God and God does not exist, again nothing is lost. However, if one chooses to believe in God and God does exist, the reward is everlasting life. On the other hand, if one chooses not to believe in God yet God does exist, the penalty is everlasting damnation. Pascal's conclusion was that belief in God provided the highest expectation.

Pascal's correspondent, Pierre Fermat, was a lawyer by trade, but a great mathematician by inclination. Born to a wealthy merchant father and a mother of nobility, Fermat studied law but developed a penchant for mathematics early in life. Married and appointed to parliament in the same year, 1631, he became the father of five children and a moderately successful civil servant. Fermat's legal career was unremarkable, although financially fruitful.

Although never trained or employed as a professional mathematician, Fermat became known, along with Pascal and Descartes, as one of the pre-eminent mathematicians of his generation. Financially secure, thanks to inheritance, marriage, and his work as a lawyer and *parlementaire*, Fermat was able to devote a large amount of free time to his studies and research in mathematics.

Fermat made important contributions in many areas of mathematics. He developed analytical geometry independent of (and actually a little bit before) Descartes, leading to an unfortunate dispute between two of the age's greatest minds. Fermat also made important discoveries in both calculus and number theory. Much of his work, however, was ignored — primarily because he did not publish during his lifetime. What was known about Fermat's mathematics was revealed only through his correspondence with other mathematicians, such as the letters he exchanged with Pascal.

No discussion of Pierre Fermat would be complete without mention of his most famous work, known today as Fermat's Last Theorem. Fermat wrote his famous words in the margin of his copy of Diophantus' work *Arithmetica*. Essentially, the conjecture Fermat read stated that the equation $x^n + y^n = z^n$ has no integer solutions x, y, z if n is an integer three or greater. Note the significance of this statement. Every schoolchild knows the Pythagorean Theorem, $x^2 + y^2 = z^2$, has many (actually infinite) solutions. In other words, if x is 3, y is 4, and z is 5, the equation is satisfied, and so on. What Fermat was reading in Diophantus' book was that there are *no* such solutions for $x^3 + y^3 = z^3$, or if the power is 4, or 5, etc.

What guaranteed the controversy which has followed Fermat's Last Theorem for over three centuries was the handwritten note Fermat made in the margin of his book: "and I have assuredly found an admirable proof of this, but the margin is too narrow to contain it." This proof, which Fermat claimed to have found, eluded mathematicians great and small, famous and not-so-famous, professional and amateur, for over 300 years. Although many tried, including such prominent mathematicians as Euler, Legendre, Dirichlet and others, none were successful. Most mathematicians came to believe that, even if Fermat really did have a proof for this theorem, it was probably not correct. This belief was supported when a proof of Fermat's Last Theorem was finally completed by Princeton mathematician Andrew Wiles in the 1990s. Wiles used mathematics that was not even in existence when Fermat wrote his famous marginal note. Wiles' proof, which he dedicated many years of diligent work to completing, was also hundreds of pages long, so Fermat's claim that "the margin is too narrow to contain it" would have been one of the great understatements in history.

The series of letters exchanged between Fermat and Pascal were brilliant, and often flawed. The two men took turns proposing solutions and then critiquing the other man's work. The following letter provides the reader a flavor of the exchange between Pascal and Fermat:

Monsieur

If I undertake to make a point with a single die in eight throws, and if we agree after the money is put at stake, that I shall not cast the first throw, it is necessary by my theory that I take ⅙ of the total sum to be impartial because of the aforesaid first throw.

And if we agree after that that I shall not play the second throw, I should, for my share, take the sixth of the remainder that is $5/36$ of the total.

If, after that, we agree that I shall not play the third throw, I should to recoup myself, take ⅙ of the remainder which is $25/216$ of the total.

And if subsequently, we agree again that I shall not cast the fourth throw, I should take ⅙ of the remainder or $125/1296$ of the total, and I agree with you that that is the value of the fourth throw supposing that one has already made the preceding plays.

But you proposed in the last example in your letter (I quote your very terms) that if I undertake to find the six in eight throws and if I have thrown three times without getting it, and if my opponent proposes that I should not play the fourth time, and if he wishes me to be justly treated, it is proper that I have $^{125}/_{1296}$ of the entire sum of our wagers.

This, however, is not true by my theory. For in this case, the three first throws having gained nothing for the player who holds the die, the total sum thus remaining at stake, he who holds the die and who agrees to not play his fourth throw should take ⅙ as his reward.

And if he has played four throws without finding the desired point and if they agree that he shall not play the fifth time, he will, nevertheless, have ⅙ of the total for his share. Since the whole sum stays in play it not only follows from the theory, but it is indeed common sense that each throw should be of equal value.

I urge you therefore (to write me) that I may know whether we agree in the theory, as I believe (we do), or whether we differ only in its application.

I am, most heartily, etc.,

Fermat.

Through this correspondence, Pascal and Fermat essentially created the modern discipline of mathematical probability. A Dutch scientist and mathematician, Christiaan Huygens — best known as the man who patented the first pendulum clock — learned of the Pascal-Fermat correspondence and quickly published the first printed version of the new ideas developing in mathematical probability in a book called *De Ratiociniis in Ludo Aleae* (*On Reasoning in Games of Chance*) in 1657.

Over the coming years, studies on the application of these new ideas to gambling appeared from some of the most important mathematicians and scientists in the world. These works often met with criticism because they were perceived as encouraging gambling. One man, Abraham de Moivre, met this head on by claiming quite the opposite in his book, *The Doctrine of Chance: A method of calculating the probabilities of events in play* (1718). Although de Moivre addressed those "who are desirous to know what foundation they go upon, when they engage in Play, whether from a motive of Gain, or barely Diversion" and offered "several practical rules" that may be useful for gamblers, he also hoped his work could "help to cure a Kind of Superstition, which has been long standing in the World, viz. that there is in Play such a thing as Luck, good or bad." In fact, in his dedication to Lord Carpenter, de Moivre maintained:

> There are many People in the World who are prepossessed with an Opinion, that the Doctrine of Chances has a Tendency to promote Play, but they soon will be undeceived.... Your Lordship does easily perceive, that this Doctrine of Chances is so far from encouraging Play, that it is rather a Guard against it, by setting in a clear Light, the Advantages and Disadvantages of those Games wherein Chance is concerned.

Like many of his contemporaries, de Moivre revealed discoveries of new ideas in probability in the process of analyzing games of chance. In particular, de Moivre defined the notion of independent events in statistics and included the first description of the normal curve. In a few centuries, these questions concerning probability and gambling would change the way scientists thought about their craft.

Pierre-Simon Laplace

Much of the early work on the theory and applications of probability was contained in an important treatise called *Analytical Theory of Probabilities* written by the brilliant

French mathematician and astronomer Pierre–Simon Laplace (1749–1827). This book served as a foundation for the study of probability for many decades to come.

Unlike many of his contemporary scientists, Laplace was not born into a wealthy and successful family. He did, however, attract the attention of wealthy patrons who helped the young Laplace enter Caen University where he planned to study theology. Those plans were abandoned when the young man found he had a talent and a love for mathematics. In fact, by the age of 19, Laplace was already chair of mathematics at the Military Academy of Paris.

One of Laplace's interests was the mathematical astronomy he found in Newton's *Principia*. In the *Principia*, Newton had been unable to account for all of the motions of the planets — this instability was literally left in God's hands as an occasional intervention was required to maintain the planets in their orbits. Laplace was able to overcome the difficulties in the Newtonian system, and in a monumental five volume work called *Celestial Mechanics,* he presented a mathematical and physical description of the solar system with every movement completely accounted for. In fact, as the story goes, when Napoleon was presented with a copy of *Celestial Mechanics*, the emperor inquired why God was never mentioned. Laplace's rather cheeky reply was: "I had no need of that hypothesis."

Laplace, who was known as the French Newton because of his work in mathematical astronomy, is also remembered for his work on the commission that designed the metric system. In addition, early in his career, Laplace published his nebular hypothesis of the solar system. In this theory, Laplace argued that our solar system (and by analogy, other solar systems) evolved from a massive ring of rotating gases. As the gases cooled, various rings broke away; upon further cooling these rings condensed to form planets. In his nebular theory, Laplace postulated that the sun was the core of the remaining gases.

If Laplace's *Analytical Theory of Probabilities* was one of the most important technical treatments of probability, his *A Philosophical Essay on Probability* was one of the most popular, since it attempted to simplify the very difficult subject matter for a general audience. In *A Philosophical Essay,* Laplace addressed the array of subjects that the application of probability could serve. He wrote that probability could help supply answers to "the important questions of life" where complete knowledge in the traditional sense was problematical. For instance, Laplace discussed the application of probability to analyzing the veracity of legal testimony. He wrote that the reliability of a testimony in a court depended on many factors: the reputation of the witness, the number of witnesses, the existence of conflicting testimonies, and whether the testimony resulted from an eyewitness account or was second-hand. However, Laplace noted, "the more extraordinary the event, the greater the need of its being supported by strong proofs." The essence of Laplace's argument was that the reliability of each testimony could be calculated.

The centerpiece of *A Philosophical Essay on Probabilities* was the section Laplace entitled "The General Principles of the Calculus of Probabilities." To begin these general principles, Laplace supplied a simple (and quite modern) definition for probability: "Probability is the ratio of the number of favorable cases to that of all the cases possible." He repeated and expanded on this definition in another section of his book:

> The theory of chance consists in reducing all the events of the same kind to a certain number of cases equally possible, that is to say, to such as we may be equally undecided about in regard to their existence, and in determining the number of cases favorable to the event

whose probability is sought. The ratio of this number to that of all the cases possible is the measure of this probability, which is thus simply a fraction whose numerator is the number of favorable cases and whose denominator is the number of all the cases possible.*

Laplace proceeded to lay down most of the known rules for calculating probabilities. These rules, for the most part, are the same rules taught in introductory probability even today: addition and product rules, dependent and independent events, and conditional probability, to name a few. Laplace also wrote about a subject that is today called mathematical expectation, but which he called mathematical hope. This is the product of the probability of an event with its ultimate value. This is, for instance, what a gambler might use to determine whether or not to play a game of chance — weighing the probability of winning against the amount to be won. It is also essentially the concept behind Pascal's Wager: even if the probability of winning (eternal life) is thought to be small, when multiplied by the enormous payout (eternal life), the expectation is high. Laplace argued that mathematical hope should be a guide for each decision made in life: "Consequently we ought always in the conduct of life to make the product of the benefit hoped for, by its probability, at least equal to the similar product relative to the loss."

In the introduction to *A Philosophical Essay on Probabilities*, Laplace explained why he attached such importance to this work on probabilities:

> I present here without the aid of analysis the principles and general results of this theory, applying them to the most important questions of life, which are indeed for the most part only problems of probability. Strictly speaking it may even be said that nearly all our knowledge is problematical; and in the small number of things which we are able to know with certainty, even in the mathematical sciences themselves, the principal means for ascertaining truth — induction and analogy — are based on probabilities; so that the entire system of human knowledge is connected with the theory set forth in this essay. Doubtless it will be seen here with interest that in considering, even in the eternal principles of reason, justice, and humanity, only the favorable chances which are constantly attached to them, there is a great advantage in following these principles and serious inconvenience in departing from them: their chances, like those favorable to lotteries, always end by prevailing in the midst of the vacillations of hazard.

Laplace reflected on the rising tide of secularism in science. Final causes (God, or some supernatural or unexplainable power) were once the goal of science, whereas now natural philosophers looked for physical laws and connecting theories:

> ALL events, even those which on account of their insignificance do not seem to follow the great laws of nature, are a result of it just as necessarily as the revolutions of the sun. In ignorance of the ties which unite such events to the entire system of the universe, they have been made to depend upon final causes or upon hazard, according as they occur and are repeated with regularity, or appear without regard to order; but these imaginary causes have gradually receded with the widening bounds of knowledge and disappear entirely before sound philosophy, which sees in them only the expression of our ignorance of the true causes.
>
> Present events are connected with preceding ones by a tie based upon the evident principle that a thing cannot occur without a cause which produces it.

*This statement of probability can be illustrated with a simple example. What is the probability of rolling an even number with one roll of a die? Since there are three ways to get an even number (2, 4, or 6), and six total outcomes (1, 2, 3, 4, 5, 6), the probability of rolling an even number is $3/6$ (or 0.5). This simple mathematical formulation for probability was key to understanding outcomes as ruled by laws, not luck.

Laplace proposed to remedy this ignorance by attaching probabilities to all phenomena in nature.

A Philosophical Essay on Probabilities is a fascinating application of probability to a variety of areas of humanity. Laplace included chapters applying probability to games of chance (gambling), natural philosophy (science), the moral (social) sciences, testimonies of witnesses, decisions of political assemblies and tribunals, as well as the mathematical treatment of all sorts of questions in probability theory. In one example, Laplace maintained that probability supported the veracity of Newton's universal law of gravitation:

> There results from this accord a probability that the flow and the ebb of the sea is due to the attraction of the sun and moon, so approaching certainty that it ought to leave room for no reasonable doubt. It changes to certainty when we consider that the attraction is derived from the law of universal gravity demonstrated by all celestial phenomena.

Laplace also made philosophical, or moral, judgments concerning the use of probability, even going so far as to tie it to happiness: "Thus in the conduct of life constant happiness is a proof of competency which should induce us to employ preferable happy persons." Laplace closed his book with a curious statement: "It is seen in this essay that the theory of probabilities is at bottom only common sense reduced to calculus."

Common sense, of course, just like happiness is very dependent upon a person's point of view. To a mathematician like Laplace, the path to happiness through the common sense approach of mathematics might seem obvious, but to many of his readers this was anything but the case. For Laplace, probability was more than just mathematics; it was a guide for the actions of a rational man: "Let us enlighten those whom we judge insufficiently instructed; but first let us examine critically our own opinions and weigh with impartiality their respective probabilities."

Statistical methods developed by Laplace and others eventually led to what has been termed "the death of certainty" in the physical sciences. For instance, three physicists, working independently in three different countries, discovered that statistical models could be used to predict the behavior of seemingly chaotic particles in samples of gas. These men, James Clerk Maxwell (1831–1879) from Scotland, Ludwig Boltzmann (1844–1906) from Austria, and Josiah Willard Gibbs (1839–1903) from the United States, made fundamental contributions to what would be called statistical mechanics. This type of uncertainty in nature that could only be predicted statistically also pervaded one of the central themes of twentieth-century physics — quantum mechanics (see chapter 12). The idea that the natural world should be understood only in terms of statistical probabilities changed the way humans viewed science. In the nineteenth century, it also began to change the way humans viewed themselves.

Francis Galton and Social Statistics

While questions in gambling initiated the mathematical study of probability, and applications to astronomy, mechanics, and other physical sciences quickly followed, statistical methods were soon being applied to the social, or moral, sciences as well. Statistical studies of mortality tables in the eighteenth century resulted in the birth of actuarial science thus allowing insurance companies, for the first time, to approach their business scientifically.

Applying statistical techniques to physical and psychological data taken from a large population of Scottish soldiers, the Belgian scientist Lambert Adolphe Jacques Quetelet (1796–1874) determined what he called his average man. Much more than simple measurements such as height and weight, Quetelet also attempted to measure such characteristics as criminal propensity by introducing what he called social physics. Quetelet grouped the measurement of each trait around the average value and produced a normal curve. He treated deviations from the average as errors. One of the results of his work was Quetelet's creation of the body mass index for measuring obesity.

If Quetelet invented social physics, it was another scientist who brought the application of probability and statistics to social questions into the mainstream. Francis Galton (1822–1911) was an English polymath in a time when such wide-ranging intellect was beginning to make way for specialization. Galton was a pioneer meteorologist, psychologist, and statistician; he applied statistics to genetics, originating the field of biometrics; he pioneered the use of fingerprinting as a unique identifier for individuals; he coined the term eugenics and advocated for the scientific study for improving the human stock; he was the founder of differential psychology; introduced the concept of correlation to statistics; and pioneered the use of questionnaires in his statistical studies of human mental capabilities.

Galton was born into a successful and intellectual Quaker family. Much like his second cousin, Charles Darwin (Erasmus Darwin was the grandfather of both), as a young man Galton found himself with plenty of time and money but not much direction. Again, like his cousin, Galton decided to travel and study the world. He mounted several expeditions into Africa and his publications made him a noted geographer while still a young man. After reading his cousin's *Origin of Species*, Galton turned his attentions to genetics and heredity. His work in these and closely related fields established Galton as one of the premier scientists of his age.

Galton's application of statistical methods to inheritance came after Mendel discovered the laws of genetics. However, since Mendel's work was for all practical purposes lost until the beginning of the nineteenth century (see chapter 14), Galton's inheritance studies did not take into account Mendelian genetics.

In an article entitled "Kinship and Correlation," Galton described traits inherited within families and explained the statistical law of correlation using examples from his work. He began by discussing one of his favorite topics, the idea of family likeness.

> In a book of mine called "Natural Inheritance," published about a year ago, I showed that the problems of family likeness fell entirely within the scope of the higher laws of chance; that we were thereby rendered capable of defining the average amount of family likeness between kinsmen in each and every degree, and of expressing the frequency with which the family likeness will depart from its average amount to any specified extent. It followed, very unfortunately for the general reader, that the problems of family likeness do not admit of being properly expressed except in the technical language of the laws of chance, and that it is impossible to discuss them adequately except through the medium of mathematics.

One of these mathematic laws developed by Galton acknowledged that offspring did not always bear the same characteristics to the same degree as the parents. Galton noticed that inherited traits showed what he termed a regression to the mean. For example, the male offspring of very tall men were often shorter than their fathers, approaching towards the mean, or average, height of all men.

Galton's explanation of correlation in heredity addressed two examples discussed in his book, *Natural Inheritance*. One was an older question pertaining to comparative anatomy, the other an exciting new application of statistics to a social question:

> One was a renewed discussion among anthropologists as to the information that the length of a particular bone — say a solitary thighbone dug out of an ancient grave — might afford concerning the stature of the unknown man to whom it belonged. It seemed to me that the anthropologists had not discussed their facts in the best statistical manner, and that they ought to have adopted a different form of treatment to any they had hitherto tried. The other circumstance arose out of the interest excited by M. Alphonse Bertillon, who proved that it was feasible to identify old criminals by an anthropometric process. The man who was suspected of having been convicted before was variously measured, and his measures were compared with those of all the criminals who had previously passed through the same process. By a contrivance analogous in principle to that on which a dictionary is constructed, the search through a register containing many tens of thousands of measures was performed with unexpected ease and precision.

Galton proceeded to discuss a method by which he attempted to analyze the two examples. He concluded:

> Reflection soon made it clear to me that not only were the two new problems identical in principle with the old one of kinship which I had already solved, but that all three of them were no more than special cases of a much more general problem — namely, that of Correlation.

Galton then attempted to explain the idea of correlation in a non-technical manner using easily understood examples. Correlation, Galton explained, does not deal with absolute measurements, but rather differences from the average. For example:

> If we were measuring statures, and had made a mark on our rule at a height equal to the average height of the race of persons whom we were considering, then it would be the distance of the top of each man's head from that mark, upward or downward as the case might be, that is wanted for our use, and not its distance upward from the ground. In speaking of the couples of brothers, and of men of the same race who were not brothers, it was the differences of stature that were noted, and not the absolute statures.

Galton also provided an example to distinguish between *relation* and *correlation*.

> A long finger usually indicates a tall person, and a tall person has usually a long finger, but by no means to-the same amount. There is relation between stature and length of finger, but no real correlation. On the other hand, the scale of variation of symmetrical limbs, such as that of the right and the left cubit, is so nearly the same that they can justly be said to be correlated.

Galton went on to explain, in a single paragraph, why a statistical view of the world was gaining so much momentum in the late nineteenth century — statistical laws seemed to apply equally to data no matter from where the data arose. Errors in astronomical observations, differences in physical characteristics, and any other example where data was found all obeyed these basic laws:

It is now beginning to be generally understood, even by merely practical statisticians, that there is truth in the theory that all variability is much of the same kind. The theory rests on the grounds that all variability is due to an uncounted number of small independent influences, acting variously in different cases. Mathematicians are able on these purely abstract grounds to develop a singularly beautiful law, known as the law of frequency of error. It is the basis of the higher statistics, and is founded upon such laws of chance as those which enable us to calculate the relative frequency of runs of luck of different lengths. The results are as precise as possible. It tells, for example, that if one-half of all the departures in a series of measures lie within 100 units of distance from the common average, three-quarters of them will lie within 171 units of distance. This kind of information is now readily to be obtained in all needed variety from well-known tables that have been calculated for the purpose, and which refer solely to what may be called the standard or the normal form of variability.

Galton finished his article by explaining other important statistical concepts — such as regression and the normal curve — in layman terms. He closed by pointing out areas of research in the social sciences that might be open to statistical analysis, namely the relationship between pauperism and crime:

There seems to be a wide field for the application of these methods to social problems. To take a possible example of such problems, I would mention the relation between pauperism and crime. I have not tried it myself; but it is easy to see that here, as in every case of relation, success would largely depend on finding quasi-normal series to deal with. Both pauperism and crime admitting of many definitions, it would be necessary to restrict the meanings of those words for the purpose of the inquiry, so that the cases to be dealt with shall be fairly homogeneous in respect to all important circumstances. To do this is the business of the statistician, who becomes assured of the soundness of his judgment in devising his restrictions when he finds that his statistics are of a quasi-normal character. If he is able to succeed in this task in the present problem, the relation between pauperism and crime would be rigorously expressed by the simple methods already explained.

Galton made bold proposals concerning what could be done with such information. If statistics could help predict the probability of pauperism, crime, and other social ills, might not society act upon this information to encourage improvement to the human stock? Galton's advocacy for eugenics made the new science a popular discipline in the late nineteenth and early twentieth century. In a paper read before the Sociological Society at a meeting at London University in 1904, Galton began by providing his definition of eugenics:

EUGENICS is the science which deals with all influences that improve the inborn qualities of a race; also with those that develop them to the utmost advantage. The improvement of the inborn qualities, or stock, of some one human population will alone be discussed here.

Galton certainly understood the possible pitfalls of eugenics. He knew that some of the definitions of what was good in a race were subjective, and he addressed these concerns:

Though no agreement could be reached as to absolute morality, the essentials of eugenics may be easily defined. All creatures would agree that it was better to be healthy than sick, vigorous than weak, well-fitted than ill-fitted for their part in life; in short, that it was better to be good rather than bad specimens of their kind, whatever that kind might be. So with men. There are a vast number of conflicting ideals, of alternative characters, of incompatible civilizations; but they are wanted to give fullness and interest to life. Society would be very dull if every man resembled the highly estimable Marcus Aurelius or Adam Bede. The aim of eugenics is to represent each class or sect by its best specimens; that done, to leave them to work out their common civilization in their own way.

However, Galton argued, the possible good for society made eugenics a practical problem for the nation:

> Let us for a moment suppose that the practice of eugenics should hereafter raise the average quality of our nation to that of its better moiety at the present day, and consider the gain. The general tone of domestic, social, and political life would be higher. The race as a whole would be less foolish, less frivolous, less excitable, and politically more provident than now. Its demagogues who "*played* to the gallery" would play to a more sensible gallery than at present. We should be better fitted to fulfill our vast imperial opportunities. Lastly, men of an order of ability which is now very rare would become more frequent, because, the level out of which they rose would itself have risen.

And finally, Galton went directly to the point. He laid out a course of action that every nation should follow to improve the value of its human stock.*

> The aim of eugenics is to bring as many influences as can be reasonably employed, to cause the useful classes in the community to contribute *more* than their proportion to the next generation. The course of procedure that lies within the functions of a learned and active society, such as the sociological may become, would be somewhat as follows:
>
> 1. Dissemination of a knowledge of the laws of heredity, so far as they are surely known, and promotion of their further study. Few seem to be aware how greatly the knowledge of what may be termed the *actuarial* side of heredity has advanced in recent years. The *average* closeness of kinship in each degree now admits of exact definition and of being treated mathematically, like birth- and death-rates, and the other topics with which actuaries are concerned.
>
> 2. Historical inquiry into the rates with which the various classes of society (classified according to civic usefulness.) have contributed to the population at various times, in ancient and modern nations. There is strong reason for believing that national rise and decline is closely connected with this influence. It seems to be the tendency of high civilization to check fertility in the upper classes,— through numerous causes, some of which are well known, others are inferred, and others again are wholly obscure. The latter class are apparently analogous to those which bar the fertility of most species of wild animals in zoological gardens. Out of the hundreds and thousands of species that have been tamed, very few indeed are fertile when their liberty is restricted and their struggles for livelihood are abolished; those which are so, and are otherwise useful to man, becoming domesticated. There is perhaps some connection between this obscure action and the disappearance of most savage races when brought into contact with high civilization, though there are other and well-known concomitant causes. But while most barbarous races disappear, some, like the negro, do not. It may therefore be expected that types of our race will be found to exist which can be highly civilized without losing fertility; nay, they may become more fertile under artificial conditions, as is the case with many domestic animals.
>
> 3. Systematic collection of facts showing the circumstances under which large and thriving families have most frequently originated; in other words, the *conditions* of eugenics.

---

*I include the following lengthy passage without editing as it illustrates Galton's thoughts on eugenics and indeed summarizes the thinking of other eugenicists of the period.

The definition of a thriving family, that will pass muster for the moment at least, is one in which the children have gained distinctly superior positions to those who were their classmates in early life. Families may be considered "*large*" that contain not less than three adult male children. It would be no great burden to a society including many members who had eugenics at heart, to initiate and to preserve a large collection of such records for the use of statistical students. The committee charged with the task would have to consider very carefully the form of their circular and the persons intrusted to distribute it. They should ask only for as much useful information as could be easily, and would be readily, supplied by any member of the family appealed to. The point to be ascertained is the *status* of the two parents at the time of their marriage, whence its more or less eugenic character might have been predicted, if the larger knowledge that we now hope to obtain had then existed. Some account would be wanted of their race, profession, and residence; also of their own respective parentages, and of their brothers and sisters. Finally the reasons would be required, why the children deserved to be entitled a "thriving" family. This manuscript collection might hereafter develop into a "*golden* book" of thriving families. The Chinese, whose customs have often much sound sense, make their honors retrospective. We might learn from them to show that respect to the parents of noteworthy children which the contributors of such valuable assets to the national wealth richly deserve. The act of systematically collecting records of thriving families would have the further advantage of familiarizing the public with the fact that eugenics had at length become a subject of serious scientific study by an energetic society.

4. Influences affecting marriage. The remarks of Lord Bacon in his essay on *Death* may appropriately be quoted here. He says with the view of minimizing its terrors: "There is no passion in the mind of men so weak but it mates and masters the fear of death.... Revenge triumphs over death; love slights it; honour aspireth to it; grief flyeth to it; fear pre-occupateth it." Exactly the same kind of considerations apply to marriage. The passion of love seems so overpowering that it may be thought folly to try to direct its course. But plain facts do not confirm this view. Social influences of all kinds have immense power in the end, and they are very various. If unsuitable marriages from the eugenic point of view were banned socially, or even regarded with the unreasonable disfavor which some attach to cousin-marriages, very few would be made. The multitude of marriage restrictions that have proved prohibitive among uncivilized people would require a volume to describe.

5. Persistence in setting forth the national importance of eugenics. There are three stages to be passed through: (1) It must be made familiar as an academic question, until its exact importance has been understood and accepted as a fact. (2) It must be recognized as a subject whose practical development deserves serious consideration. (3) It must be introduced into the national conscience, like a new religion. It has, indeed, strong claims to become an orthodox religious, tenet of the future, for eugenics co–operate with the workings of nature by securing that humanity shall be represented by the fittest races. What nature does blindly, slowly, and ruthlessly, man may do providently, quickly, and kindly. As it lies within his power, so it becomes his duty to work in that direction. The improvement of our stock seems to me one of the highest objects that we can reasonably attempt. We are ignorant of the ultimate destinies of humanity, but feel perfectly sure that it is as noble a work to raise its level, in the sense already explained, as it would be disgraceful to abase it. I see no impossibility in eugenics becoming a religious dogma among mankind, but its details must first be worked out sedulously in the study. Overzeal leading to hasty action would do harm, by holding out expectations of a near golden age, which will certainly be falsified and cause the

science to be discredited. The first and main point is to secure the general intellectual acceptance of eugenics as a hopeful and most important study. Then let its principles work into the heart of the nation, which will gradually give practical effect to them in ways that we may not wholly foresee.

Of course, Galton's call to action on eugenics revealed many underlying difficulties: racist undertones, marriage and reproduction restrictions, and other socially divisive issues made eugenics a lightning rod for disputes.

In fact, after Galton read his paper to the Sociological Society, a debate ensued among the members present. In attendance was an eclectic mix of English scientists (Karl Pearson, for instance) and physicians, literary figures such as H.G. Wells and George Bernard Shaw, and various other intellectuals interested in the social sciences. Most in attendance generally supported Galton's ideas, but there were criticisms. A Dr. Maudsley cautioned:

> In considering the question of hereditary influences, as I have done for some long period of my life, one met with the difficulty, which must have occurred to everyone here, that in any family of which you take cognizance you may find one member, a son, like his mother or father, or like a mixture of the two, or more like his mother, or that he harks back to some distant ancestor; and then again you will find one not in the least like father or mother or any relatives, so far as you know. There is a variation, or whatever you may call it, of which in our present knowledge you cannot give the least explanation.... From my long experience as a physician I could give instances in every department — in science, in literature, in art — in which one member of the family has risen to extraordinary prominence, almost genius perhaps, and another has suffered from mental disorder.
>
> Now, how can we account for these facts on any of the known data on which we have at present to rely? In my opinion, we shall have to go far deeper down than we have been able to go by any present means of observation — to the corpuscles, atoms, electrons, or whatever else there may be; and we shall find these subjected to subtle influences of mind and body during their formations and combinations, of which we hardly realize the importance....
>
> In view of these difficulties of the subject, it has always seemed to me that we must not be hasty in coming to conclusions and laying down any rules for the breeding of humans and the development of a eugenic conscience.

Dr. Mercier agreed, adding

> Mr. Galton speaks of the laws of heredity, and dissemination of a knowledge of the laws of heredity in so far as we know them, and the qualification is very necessary. For, in so far as we know the laws, they are so obscure and complex that to us they work out as chance. We cannot detect any practical difference in the working of the laws of heredity and the way in which dice may be taken out of a lucky bag. It is quite impossible to predict from the constitution of the parents what the constitution of the offspring is going to be, even in the remotest degree.

Professor Raphael Weldon, on the other hand, was perhaps Galton's most ardent defender at the meeting. Weldon was a zoologist with an interest in statistical heredity. He became (along with Karl Pearson) the first editor of the journal *Biometrika*. After Galton's reading to the Sociological Society, Weldon addressed two criticisms often directed at Galton and eugenics:

> There are two sets of objections which have been used against the points made by Dr. Galton: One set criticises the statistical method on the ground that it cannot account for a num-

ber of phenomena.... I venture to say it is no proper part of statistics to account for anything, but it is the triumph of statistics that it can describe, and with a very fair degree of accuracy, a large number of phenomena. And, as I conceive the matter, the essential object of eugenics is not to put forward any theory of causation of hereditary phenomena; it is to diffuse the knowledge of what these phenomena really are. We may not be able to account for the formation of a Shakespeare, but we may be able to tabulate a scheme of inheritance which will indicate with very fair accuracy, the percentage of cases in which children of exceptional ability result from a particular type of marriage.

The second objection to Galton's work addressed by Weldon acknowledged the newly discovered work of Mendel:

Because a large number of apparently simple results have been attained in experimental breeding establishments, and especially by the Austrian abbot, Gregor Mendel, it has been too lightly assumed that these phenomena have henceforward superseded the actuarial method, and that the only reliable method is experiment on simple characters, such as those initiated by Mr. Mendel and carried out by Mr. Bateson in England, in Holland by Professor Defries, and by an increasing number of men all over Europe. But the statistical method is itself necessary in order to test the results of the experiments which are supposed to supersede it. The question whether there is really an agreement between experience and hypothesis is in nearly every case hard to answer, and can be achieved only by the use of this actuarial method which Mr. Galton has taught us to apply to biological problems.

As the invited speaker, Galton was given the last word. Evidently, he was not pleased with the majority of the debate, singling out several speakers who "really seemed to me to be living forty years ago; they displayed so little knowledge of what has been done since" and others who "were really not acquainted with the facts, and they ought not to have spoken at all." Galton closed with:

I have little more to say, except that I do feel that if the society is to do any good work in this direction, it must attack it in a much better way than the majority of speakers seem to have done tonight.

In spite of his disappointment in those gathered at the meeting, Galton's work did influence a generation of social scientists. Karl Pearson, one of the scientists present at the meeting, became one of the leaders in the new science of biometrics. And eugenics, for a time, was studied and taught all over the world. Unfortunately, abuses in the name of eugenics began to slowly erode its credibility. Sterilization laws for criminals; laws encouraging immigration from nations whose citizens exhibited desirable traits and prohibiting or sharply regulating immigrants from undesirable nations; and finally, the atrocities of fascist regimes in the name of ethnic purity, all led to the eventual demise of eugenics as an accepted science. Although eugenics is one result of statistics in social sciences that did not stand the test of time, statistical and probabilistic analysis in modern branches of the social sciences is a lasting legacy of Galton and his contemporaries.

PRIMARY SOURCES

Fermat, Pierre. "Letter from Fermat to Pascal." *A Source Book in Mathematics*, edited by David Eugene Smith. New York: McGraw-Hill, 1929.

Galton, Francis. "Eugenics: Its Definition, Scope, and Aims." *The American Journal of Sociology* X, no. 1 (1904).

_____."Kinship and Correlation." *North American Review* 150 (1890). www.galton.org.

Laplace, Pierre. *A Philosophical Essay on Probability*, translated by F.W. Truscott and F.L. Emory. New York: John Wiley and Sons; London: Chapman and Hall, Limited, 1902.

Moivre, Abraham de. *The Doctrine of Chance: A Method of Calculating the Probabilities of Events in Play*. London, 1718.

OTHER SOURCES

Bulmer, M.G. *Francis Galton: Pioneer of Heredity and Biometry.* Baltimore: Johns Hopkins University Press, 2003.

Cowan, Ruth Schwartz. *Sir Francis Galton and the Study of Heredity in the Nineteenth Century.* New York: Garland, 1985.

Daston, Lorraine. *Classical Probability in the Enlightenment.* Princeton: Princeton University Press, 1988.

David, F.N. *Games, Gods, and Gambling.* New York: Oxford University Press, 1962.

Devlin, Keith J. *The Unfinished Game: Pascal, Fermat, and the Seventeenth-Century Letter That Made the World Modern.* New York: Basic Books, 2008.

Gigerenzer, Gerd, Zeno Swijtink, Theodore M. Porter, Lorraine Daston, J. Beatty, and Lorenz Krüger, eds. *The Empire of Chance: How Probability Changed Science and Everyday Life.* Cambridge: Cambridge University Press, 1989.

Gillham, Nicholas W. *A Life of Sir Francis Galton: From African Exploration to the Birth of Eugenics* New York: Oxford University Press, 2001.

Gillispie, Charles Coulston, Robert Fox, and I. Grattan-Guinness. *Pierre-Simon Laplace, 1749–1827: A Life in Exact Science.* Princeton: Princeton University Press, 1997.

Hacking, Ian. *The Emergence of Probability.* London: Cambridge University Press, 1975.

Hahn, Roger. *Pierre Simon Laplace, 1749–1827: A Determined Scientist.* Cambridge: Harvard University Press, 2005.

Hald, Anders. *A History of Probability and Statistics and Their Applications Before 1750.* New York: Wiley, 1990.

Hazelton, Roger. *Blaise Pascal: The Genius of His Thought.* Philadelphia: Westminster Press, 1974.

Krüger, Lorenz, Lorraine J. Daston, and Michael Heidelberger, eds. *The Probabilistic Revolution.* 2 vols. Cambridge: MIT Press, 1987.

Mahoney, Michael Sean. *The Mathematical Career of Pierre de Fermat.* Princeton: Princeton University Press, 1973.

Owen, D.B. *On the History of Statistics and Probability.* New York: Dekker, 1976.

Porter, Theodore M. *The Rise of Statistical Thinking, 1820–1900.* Princeton: Princeton University Press, 1986.

Stigler, Stephen M. *The History of statistics: The Measurement of Uncertainty Before 1900.* Cambridge: The Belknap Press of Harvard University Press, 1986.

CHAPTER 12

Einstein, Bohr, and
Twentieth-Century Physics

In any molecular system consisting of positive nuclei and electrons in which the nuclei are at rest relative to each other and the electrons move in circular orbits, the angular momentum of every electron round the centre of its orbit will in the permanent state of the system be equal to $h/2\pi$ where h is Planck's constant — Niels Bohr, "On the Constitution of Atoms and Molecules" (1913)

Before the advent of relativity, physics recognized two conservation laws of fundamental importance, namely, the law of the conservation of energy and the law of the conservation of mass; these two fundamental laws appeared to be quite independent of each other. By means of the theory of relativity they have been united into one law.— Albert Einstein, *Relativity: The Special and General Theory* (1920)

In a speech given at the 1900 meeting of the British Association for the Advancement of Science, William Thomson (better known as Lord Kelvin), infamously stated: "There is nothing new to be discovered in physics now. All that remains is more and more precise measurement." Ironically, physics was on the brink of a revolution that would rival any in the history of science. Over the next several decades, a new and surprising view of the atom emerged. Along with this new conception of the structure of the atom, an extraordinary concept called quantum theory turned modern physics upside down. Finally, another revolutionary theory, relativity, called into question everything that physicists thought they understood about the universe. Although a diverse group of scientists from around the world contributed to these new ideas in physics, two—Albert Einstein and Niels Bohr—stand out for their creative and imaginative work. Before Einstein and Bohr would change the world, however, the work of other physicists and chemists paved a path for understanding the fundamental nature of matter.

Fundamental Particles

John Dalton's atomic theory (see chapter 8) treated the atom as a solid sphere — a "billiard ball." The word atom, after all, comes from the Greek for indivisible. Yet, even by the turn of the twentieth century, discoveries were calling into question this traditional view of the atom. In the last years of the nineteenth century a young English physicist working at the prestigious Cavendish Laboratory in Cambridge performed several notable experiments that led him to conclude that the atom was not indivisible. J.J. Thomson showed that the mysterious emissions known as cathode rays were bent by an electric field. Thomson concluded that these rays, which he called corpuscles, were negatively charged components of the atom. In his Nobel lecture (1906), Thomson described his experiments and subsequent interpretation of the results:

> In this lecture I wish to give an account of some investigations which have led to the conclusion that the carriers of negative electricity are bodies, which I have called corpuscles, having a mass very much smaller than that of the atom of any known element, and are of the same character from whatever source the negative electricity may be derived.

Thomson elaborated further on the nature of cathode rays:

> Two views were prevalent: one, which was chiefly supported by English physicists, was that the rays are negatively electrified bodies shot off from the cathode with great velocity; the other view, which was held by the great majority of German physicists, was that the rays are some kind of ethereal vibration or waves. The arguments in favor of the rays being negatively charged particles are primarily that they are deflected by a magnet in just the same way as moving, negatively electrified particles.
>
> ... The next step in the proof that cathode rays are negatively charged particles was to show that when they are caught in a metal vessel they give up to it a charge of negative electricity. This was first done by Perrin.
>
> ... If the rays are charged with negative electricity they ought to be deflected by an electrified body as well as by a magnet. In the earlier experiments made on this point no such deflection was observed. The reason of this has been shown to be that when cathode rays pass through a gas they make it a conductor of electricity, so that if there is any appreciable quantity of gas in the vessel through which the rays are passing, this gas will become a conductor of electricity and the rays will be surrounded by a conductor which will screen them from the effect of electric force, just as the metal covering of an electroscope screens off all external electric effects. By exhausting the vacuum tube until there was only an exceedingly small quantity of air left in to be made a conductor, I was able to get rid of this effect and to obtain the electric deflection of the cathode rays. This deflection had a direction which indicated a negative charge on the rays. Thus, cathode rays are deflected by both magnetic and electric forces, just as negatively electrified particles would be.

After carefully calculating the ratio of the particle's charge to its mass, Thomson concluded that this corpuscle is extremely small, even compared to the atom, "hence we are driven to the conclusion that the mass of the corpuscle is only about $\frac{1}{1,700}$ of that of the hydrogen atom. Thus the atom is not the ultimate limit to the subdivision of matter."

The atom is not the smallest particle in nature! It is composed of other, smaller particles, in particular this corpuscle, which was soon to be called the electron.

The discovery of the electron changed our understanding of the fundamental makeup of the atom. If not a dense, hard, indivisible sphere, what, then was an atom? Thomson proposed that the atom was a sphere of positively charged fluid with very small negatively charged particles interspersed. This model became known as the plum pudding model, with the positively charged fluid being the pudding while the electrons were the plums mixed throughout. Thomson's model of the atom underwent a major revision at the hands of another physicist who worked under him briefly at the Cavendish Laboratory.

In 1895, a young New Zealander by the name of Ernest Rutherford arrived at Cambridge. Rutherford collaborated with Thomson before moving on to Canada where he did groundbreaking work on radioactivity at McGill University. It was for this work that Rutherford was later awarded the Nobel Prize in Chemistry. Returning to England in 1907, Rutherford was appointed professor of physics at Manchester University. Then, twelve years later, the New Zealander succeeded his mentor, Thomson, as director of the Cavendish Laboratory at Cambridge.

While at Manchester, Rutherford conceived of one of the most famous experiments in the history of science. Rutherford directed Hans Geiger and Ernest Marsden to aim a beam of alpha particles at a sheet of thin gold foil surrounded by a detector that indicated when it was struck by an alpha particle. (The alpha particle, a product of radioactive decay, is now known to be a helium nucleus). The results of the experiment shocked Rutherford and his collaborators. Thomson's plum pudding model predicted that most of the alpha particles would pass through the gold foil unabated, while a few would be deflected slightly. Instead, Rutherford and his team noted that a few of the particles scattered over great angles, some even deflecting back in the direction from whence they came. Rutherford colorfully described the astonishing results: "It was quite the most incredible event that has ever happened to me in my life. It was almost as incredible as if you fired a 15-inch shell at a piece of tissue paper and it came back and hit you."

The gold foil experiment led Rutherford to a new theory concerning the structure of the atom. Rutherford explained the results of the experiment by proposing that atoms contained an extremely small, dense, positively-charged center (the nucleus) surrounded by electrons in orbit around the center. Since the electrons orbited the nucleus much like the planets orbit the sun, Rutherford's theory became known as the planetary model of the atom.

When English physicist James Chadwick discovered the neutron in 1932, it seemed that the picture of the atom was complete: a tiny, hard, nucleus composed of protons and neutrons surrounded by orbiting electrons in wide-open spaces. Of course, this was not the case. Throughout the remainder of the twentieth century scientists continued to discover smaller (and stranger) sub-atomic particles. Quarks, leptons, and bosons entered the lexicon as building blocks of matter even more basic than protons, neutrons and electrons. Many new particles were discovered (and some even created) by an ingenious invention called the cyclotron. The cyclotron is a type of particle accelerator which uses a magnetic field to accelerate a charged particle — such as a proton — to very high speeds. These charged particles are then collided with various materials, often destroying the target and in the process revealing new particles. The first cyclotron was built by American physicist Ernest Lawrence in 1929. Over the ensuing decades, new and more powerful particle accelerators helped scientists to unlock many of the secrets of nature. The Large

Hadron Collider, the most powerful particle accelerator ever built, became fully operational in 2010. With this massive (and expensive) machine, scientists hope to delve deeper into the mysteries of subatomic particles.

Ernest Rutherford's planetary model was widely accepted until Niels Bohr introduced innovative ideas in quantum theory to once more revise scientists' understanding of the structure of the atom. In the early decades of the twentieth century, quantum theory became one of the most important new ideas in physics. The best minds in the world, from Bohr to Einstein, made contributions to quantum theory; however, this conceptually complex theory also stirred dissent among scientists as they searched for concrete solutions to increasingly abstract problems.

The Birth of Quantum Theory

Perhaps the greatest triumph of the Enlightenment era was the certainty that Newton's laws provided. This certainty implied a deterministic universe; philosophers believed that the world could be understood — exactly and completely — through empiricism and mathematics. One of the fundamental revolutions in science began in the latter half of the nineteenth century when scientists from various disciplines began discovering characteristics of nature that could not be described exactly and completely. Instead, they found that nature behaved unpredictably and therefore statistical methods were required to describe the observed phenomena.

Of course, one of the earliest theories to introduce an element of chance into science was Darwin's natural selection. Evolution, according to Darwin, was not deterministic because chance mutations were the driving force behind natural selection. In the physical sciences, two physicists — Ludwig Boltzmann and James Clerk Maxwell — found that gases behave in a way which can only be described statistically. Their discovery fundamentally changed scientists' approach to the study of mechanics and thermodynamics. Perhaps the biggest blow to determinism, however, came from a paper published the same year as Kelvin's dire statement about the end of new discoveries in physics. This paper, written by the German physicist Max Planck, made a radical new claim concerning the release of energy in natural systems.

In 1900, Max Planck (1858–1947), professor of physics at the University of Berlin, proposed that energy was not continuous, but rather emitted in discrete packets, or quanta. His research led him to a formula which related energy to the frequency of the radiation using a constant, h, now called Planck's constant. At the time, and for many years to come, Planck and other scientists struggled with the physical interpretation of the quantum theory. Planck wrote:

> Nothing can as yet be said with certainty of the dynamical representation of such quanta. Perhaps one could imagine quanta occurring in this manner, viz. that any source of radiation can only emit energy after the energy has reached a certain value, as, for example, a rubber tube, into which air is gradually pumped, suddenly bursts, and discharges its contents when a certain definite quantity has been pumped in.
>
> ... In all cases, the quantum hypothesis has given rise to this idea, that in Nature, changes occur which are not continuous, but of an explosive nature. I need only mention that this

idea has been brought into prominence by the discovery of, and closer research into, radio-active phenomena. The difficulties connected with exact investigations are lessened, since the results obtained on the quantum hypotheses agree better with observation than do those deduced from all previous theories.

In his Nobel lecture of 1918, Planck reflected upon the difficulties presented by a theory that seemed counterintuitive and conflicted directly with classical physics:

When I look back to the time, already twenty years ago, when the concept and magnitude of the physical quantum of action began, for the first time, to unfold from the mass of experimental facts, and again, to the long and ever tortuous path which led, finally, to its disclosure, the whole development seems to me to provide a fresh illustration of the long-since proved saying of Goethe's that man errs as long as he strives.

... Either the quantum of action was a fictional quantity, then the whole deduction of the radiation law was in the main illusory and represented nothing more than an empty non-significant play on formulae, or the derivation of the radiation law was based on a sound physical conception. In this case the quantum of action must play a fundamental role in physics, and here was something entirely new, never before heard of, which seemed called upon to basically revise all our physical thinking, built as this was, since the establishment of the infinitesimal calculus by Leibniz and Newton, upon the acceptance of the continuity of all causative connections.... Experiment has decided for the second alternative.

Even twenty years after his initial work, Planck admitted that the theory was yet incomplete. However, he also confidently asserted that only a matter of time would show the theory to be correct.

To be sure, the introduction of the quantum of action has not yet produced a genuine quantum theory.... But numbers decide, and the result is that the roles, compared with earlier times, have gradually changed. What initially was a problem of fitting a new and strange element, with more or less gentle pressure, into what was generally regarded as a fixed frame has become a question of coping with an intruder who, after appropriating an assured place, has gone over to the offensive; and today it has become obvious that the old framework must somehow or other be burst asunder. It is merely a question of where and to what degree.

Although Max Planck was considered a leader in German science for decades to come, he later took a position against what became known as the Copenhagen Interpretation of quantum theory. This interpretation of his own work, championed by Niels Bohr among others, became the widely accepted view in spite of Planck's opposition.

Niels Bohr, the Atom, and Quantum Mechanics

Perhaps the second most famous scientist of the twentieth century (after Albert Einstein) was the Danish physicist Niels Bohr (1885–1962). Niels' father was a professor of physiology and his mother came from a wealthy Jewish family, so Niels and his brother and sister were raised in a comfortable and intellectually stimulating environment. Niels was extremely close to his brother, Harald, who became a well-respected mathematician. While still an undergraduate at the University of Copenhagen, Niels Bohr won a national prize for a paper he wrote describing an experiment carried out in his father's laboratory. In 1911, Niels was awarded a doctorate in physics from Copenhagen. Like Ernest Rutherford before him — and many other important physicists of the era — Bohr worked under J.J. Thomson at the prestigious Cavendish Laboratory in Cambridge.

Soon Bohr was offered the opportunity to journey to Manchester and work with Rutherford — an opportunity that he quickly accepted. It was at Manchester that Bohr, taking Rutherford's planetary model as a starting point, conceived of a new theory of the structure of the atom. Bohr described Rutherford's atomic model:

> In order to explain the results of experiments on scattering of d-rays by matter Prof. Rutherford has given a theory of the structure of atoms. According to this theory, the atom consists of a positively charged nucleus surrounded by a system of electrons kept together by attractive forces from the nucleus; the total negative charge of the electrons is equal to the positive charge of the nucleus. Further, the nucleus is assumed to be the seat of the essential part of the mass of the atom, and to have linear dimensions exceedingly small compared with the linear dimensions of the whole atom. The number of electrons in an atom is deduced to be approximately equal to half the atomic weight. Great interest is to be attributed to this atom-model; for, as Rutherford has shown, the assumption of the existence of nuclei, as those in question, seems to be necessary in order to account for the results of the experiments on large angle scattering of the rays.

Bohr proceeded to introduce Planck's quantum theory into the mix by postulating that the electrons orbit in only specific discrete energy states. When an electron jumps from one energy state to another, Bohr continued, a photon is either emitted or absorbed and the energy of this jump is governed by a formula involving Planck's constant. Bohr explained the need for incorporating quantum theory into the structure of the atom:

> The inadequacy of the classical electrodynamics in accounting for the properties of atoms from an atom-model as Rutherford's, will appear very clearly if we consider a simple system consisting of a positively charged nucleus of very small dimensions and an electron describing closed orbits around it.
> ... Now the essential point in Planck's theory of radiation is that the energy radiation from an atomic system does not take place in the continuous way assumed in the ordinary electrodynamics, but that it, on the contrary, takes place in distinctly separated emissions....
> ... In any molecular system consisting of positive nuclei and electrons in which the nuclei are at rest relative to each other and the electrons move in circular orbits, the angular momentum of every electron round the centre of its orbit will in the permanent state of the system be equal to where h is Planck's constant.

It was for this work on the structure of the atom that Bohr was awarded the Nobel Prize in 1922.

Bohr returned to his native Denmark already an internationally known physicist. He became the founding director of the Institute of Theoretical Physics at the University of Copenhagen, a post from which he directed a long line of young physicists who came to work with him. One of these talented scientists was a young German physicist, Werner Heisenberg. While working with Bohr at Copenhagen, Heisenberg made important advances in quantum theory, developing advanced mathematical tools that resulted in new directions for understanding quantum mechanics.

As Heisenberg continued his work in quantum mechanics, he began to uncover a most perplexing characteristic of particle motion. Heisenberg discovered that attempting to measure both the momentum and position of a particle simultaneously led to uncertainties in the measurements. Heisenberg's Uncertainty Principle was controversial, but eventually his work, combined with the work of Bohr, Wolfgang Pauli, Max Born, and others, led to the Copenhagen Interpretation of quantum mechanics and became widely accepted among physicists (not without a few major dissenters) as the best interpretation of quantum mechanics.

By the mid–1930s, Niels Bohr had become an icon of science throughout the world. In 1940, however, the Nazi occupation of Denmark signaled an ominous change for the Bohr family. His mother's Jewish heritage made life in Denmark increasingly more perilous, until finally Bohr made a daring nighttime escape to Sweden on a fishing boat. From there, he made his way to London, and eventually to the United States where his brilliance was immediately put to use on the Manhattan Project, a top-secret project to build the world's first atomic bomb. Interestingly, Werner Heisenberg, Bohr's former protégée in Copenhagen, remained in Germany and directed Germany's own atomic bomb research. In 1941, before Bohr was forced to flee Denmark, Heisenberg traveled to Copenhagen to see his mentor. The purpose and outcome of this meeting have been debated by historians, primarily because there are conflicting reports from the parties involved. Did Heisenberg come to Bohr to warn him (and by extension, the world) of the German atomic bomb project? Did Heisenberg and some of his fellow physicists purposely slow down — or even sabotage — Hitler's efforts to build a super weapon? Whatever the truth behind the meeting, after the war ended Allied forces found a German atomic bomb project in a much less advanced stage than Allied scientists and politicians had feared.

Bohr returned to his native Copenhagen after the war, where he continued to work in physics. He also became an activist for the peaceful use of atomic energy and an advocate for sharing nuclear technologies among countries as a means for ensuring international cooperation (a passion shared by the other elder statesman of science, Albert Einstein). Bohr and his wife, Margrethe, had six sons, four of whom survived to adulthood. One son, Aage, became an important physicist in his own right, working with his father on the Manhattan Project and winning his own Nobel Prize for physics in 1975.

Albert Einstein — Twentieth Century Icon of Science

Born in Ulm, Germany, in 1879, the family of young Albert Einstein moved often. As a boy, Albert was educated in Munich, and then in Switzerland. He graduated from the Swiss Federal Polytechnic School in 1901, and accepted a job with the Swiss patent office after unsuccessfully seeking a teaching position. While employed for seven years as a technical expert in the Swiss patent office, Einstein developed many of the theories that soon led to acclaim and world-wide fame. In 1905 alone, this unknown, low-level bureaucrat from Zurich published four revolutionary papers. Two of these papers made important contributions to the fledgling fields of quantum mechanics and statistical mechanics. The third, in an attempt to unify classical mechanics with electrodynamics, introduced Einstein's special theory of relativity. The fourth and final of paper of Einstein's *Annus Mirabilis* (Miracle Year) showed the equivalence of matter and energy and introduced his famous formula, $E = mc^2$. Rounding out a remarkable year for Einstein, he was granted a doctorate in physics from the University of Zurich in 1905.

Einstein finally began to reap the rewards of his brilliant work when he was appointed as a lecturer in physics at the University of Bern (Switzerland) in 1908. The following year, Einstein won an appointment as professor of physics at the University of Zurich. As his reputation grew, Einstein moved often from position to position. In 1911 he was offered and accepted a position as professor of theoretical physics at Prague, a position he held for only one year before returning to Zurich at the prestigious Eidgenössische Technische Hochschule. Finally, in 1914, Einstein moved back to his native Germany where he accepted an appointment as chair of theoretical physics at the University of Berlin, as well as director of the newly established Kaiser Wilhelm Institute of Physics.

Einstein began to feel some effects of anti–Semitism in Germany as early as 1920. Although he remained in Berlin until 1932, his growing fame led to many offers from universities all over the world. Einstein finally accepted an offer to join the Institute of Advanced Study (IAS) on the campus of Princeton University, leaving Germany permanently just before the Nazis took power. Einstein remained at the IAS until his death over two decades later, working with some of the most brilliant scientific minds in the world. Einstein became a United States citizen in 1940 and lived the remainder of his life as the world's (and perhaps history's) most renowned scientist. He was given numerous honors from around the world, and was even offered the presidency of Israel, which he refused. Einstein married twice. The first marriage resulted in a divorce, and the second, to his cousin Elsa, lasted until her death in 1936. Albert Einstein, the iconic scientific figure of the twentieth century, died in 1955 in Princeton, New Jersey.

Einstein's Physics

The first paper published by Einstein in 1905 addressed the photoelectric effect, which not only provided solid evidence for Planck's quantum theory but also laid the groundwork for future understanding of the dual nature of light (the idea that light could sometimes be understood as behaving like a wave, but in other situations it behaves more like a stream of particles). It was this work on the photoelectric effect, not relativity, that garnered Einstein a Nobel Prize in 1921. Later in 1905, Einstein published his research into the irregular motion of particles suspended in a liquid. This investigation into Brownian motion, as it is called, provided some of the earliest physical evidence of the actual existence of molecules.

The name Albert Einstein is most universally associated with the theory of relativity. However, Einstein actually developed two theories of relativity. The first, which he published in 1905, is the *special* theory of relativity. Special relativity is based on observers in uniform relative motion, whereas the *general* theory of relativity — published in 1916 — extended the theory to observers in relative acceleration and revolutionized how scientists think about gravity.

In his *Relativity: The Special and General Theory*, Einstein attempted to explain his conclusions in such a way that people who "are interested in the theory, but who are not conversant with the mathematical apparatus of theoretical physics" might understand. He began his explanation of the special theory of relativity by addressing the difficulties in understanding seemingly simple terms such as space and time. First, what does it mean for an object to move through space?

> The purpose of mechanics is to describe how bodies change their position in space with "time." I should load my conscience with grave sins against the sacred spirit of lucidity were I to formulate the aims of mechanics in this way, without serious reflection and detailed explanations. Let us proceed to disclose these sins.
>
> ... It is not clear what is to be understood here by "position" and "space." I stand at the window of a railway carriage which is travelling uniformly, and drop a stone on the embankment, without throwing it. Then, disregarding the influence of the air resistance, I see the stone descend in a straight line. A pedestrian who observes the misdeed from the footpath notices that the stone falls to earth in a parabolic curve. I now ask: Do the "positions" traversed by the stone lie "in reality" on a straight line or on a parabola? Moreover, what is meant here by motion "in space"? From the considerations of the previous section the answer is self-evident. In the first place we entirely shun the vague word "space," of which, we must honestly acknowledge, we cannot form the slightest conception, and we replace it by "motion relative to a practically rigid body of reference." The positions relative to the body of reference (railway carriage or embankment) have already been defined in detail in the preceding section. If instead of "body of reference" we insert "system of co–ordinates," which is a useful idea for mathematical description, we are in a position to say: The stone traverses a straight line relative to a system of co–ordinates rigidly attached to the carriage, but relative to a system of co–ordinates rigidly attached to the ground (embankment) it describes a parabola. With the aid of this example it is clearly seen that there is no such thing as an independently existing trajectory..., but only a trajectory relative to a particular body of reference.

Time, Einstein continued, is equally problematic. Once again, he invokes the train traveling along an embankment to explain:

> Lightning has struck the rails on our railway embankment at two places A and B far distant from each other. I make the additional assertion that these two lightning flashes occurred simultaneously. If I ask you whether there is sense in this statement, you will answer my question with a decided "Yes." But if I now approach you with the request to explain to me the sense of the statement more precisely, you find after some consideration that the answer to this question is not so easy as it appears at first sight.

First, Einstein suggested, consider a simple and straightforward answer to the question, "Did the lightning flashes happen simultaneously?"

> After thinking the matter over for some time you then offer the following suggestion with which to test simultaneity. By measuring along the rails, the connecting line AB should be measured up and an observer placed at the mid-point M of the distance AB. This observer should be supplied with an arrangement (e.g. two mirrors inclined at) which allows him visually to observe both places A and B at the same time. If the observer perceives the two flashes of lightning at the same time, then they are simultaneous.

After addressing several difficulties in measuring time and understanding the idea of simultaneity, Einstein proposed this thought experiment:

> We suppose a very long train travelling along the rails with the constant velocity v and in the direction indicated in Fig 1. People travelling in this train will with a vantage view the train as a rigid reference-body (co–ordinate system); they regard all events in reference to the train.

Then every event which takes place along the line also takes place at a particular point of the train. Also the definition of simultaneity can be given relative to the train in exactly the same way as with respect to the embankment. As a natural consequence, however, the following question arises. Are two events (e.g. the two strokes of lightning A and B) which are simultaneous with reference to the railway embankment also simultaneous relatively to the train? We shall show directly that the answer must be in the negative.

... When we say that the lightning strokes A and B are simultaneous with respect to the embankment, we mean: the rays of light emitted at the places A and B, where the lightning occurs, meet each other at the mid-point M of the length A arrow B of the embankment. But the events A and B also correspond to positions A and B on the train. Let M1 be the mid-point of the distance A arrow B on the travelling train. Just when the flashes (as judged from the embankment) of lightning occur, this point M1 naturally coincides with the point M but it moves towards the right in the diagram with the velocity v of the train. If an observer sitting in the position M1 in the train did not possess this velocity, then he would remain permanently at M, and the light rays emitted by the flashes of lightning A and B would reach him simultaneously, i.e. they would meet just where he is situated. Now in reality (considered with reference to the railway embankment) he is hastening towards the beam of light coming from B, whilst he is riding on ahead of the beam of light coming from A. Hence the observer will see the beam of light emitted from B earlier than he will see that emitted from A. Observers who take the railway train as their reference-body must therefore come to the conclusion that the lightning flash B took place earlier than the lightning flash A. We thus arrive at the important result: Events which are simultaneous with reference to the embankment are not simultaneous with respect to the train, and vice versa (relativity of simultaneity). Every reference-body (co–ordinate system) has its own particular time; unless we are told the reference-body to which the statement of time refers, there is no meaning in a statement of the time of an event.

Einstein's conclusion, that space and time are relative to the observer's coordinate system, was in direct contrast to traditional (Newtonian) mechanics which held that space and time were absolute:

Now before the advent of the theory of relativity it had always tacitly been assumed in physics that the statement of time had an absolute significance, i.e. that it is independent of the state of motion of the body of reference. But we have just seen that this assumption is incompatible with the most natural definition of simultaneity.

And finally, Einstein concluded his argument by combining the notions of mass and energy into one law:

The most important result of a general character to which the special theory of relativity has led is concerned with the conception of mass. Before the advent of relativity, physics recognized two conservation laws of fundamental importance, namely, the law of the conservation of energy and the law of the conservation of mass these two fundamental laws appeared to be quite independent of each other. By means of the theory of relativity they have been united into one law.

That law, of course, is the iconic $E = mc^2$.

After offering his explanation of the special theory of relativity, Einstein proceeded to the general theory of relativity. In a discussion of the bending of light by a gravitational field as a consequence of relativity, Einstein proposed an observational experiment to confirm his theory:

As seen from the earth, certain fixed stars appear to be in the neighborhood of the sun, and are thus capable of observation during a total eclipse of the sun. At such times, these stars ought to appear to be displaced outwards from the sun by an amount indicated above, as compared with their apparent position in the sky when the sun is situated at another part of the heavens.

In 1919, the British astrophysicist Arthur Eddington led an expedition to Africa to observe an eclipse of the sun. Eddington's observations and calculations confirmed Einstein's prediction and supplied the first physical evidence of the veracity of the theory of relativity.

The Dispute Between Bohr and Einstein

Paradoxically, although Einstein played a central role in the early development of quantum theory, he did not agree with the direction the theory took in the decades to follow. In particular, the great German physicist was reluctant to accept the indeterminate nature of quantum mechanics, and more than once was known to utter the phrase, "God does not play dice with the universe." This put Einstein at odds with Bohr and many of the other leading physicists of the era.

In a volume of collected works honoring Einstein entitled *Albert Einstein: Philosopher-Scientist*, Bohr wrote of the differences between Einstein and himself. However, Bohr began by crediting Einstein with an important early contribution to quantum theory in the context of Planck's original ideas:

> While, in his work, Planck was principally concerned with considerations of essentially statistical character and with great caution refrained from definite conclusions as to the extent to which the existence of the quantum implied a departure from the foundations of mechanics and electrodynamics. Einstein's great original contribution to Quantum Theory (1905) was just the recognition of how physical phenomena like the photo-effect may depend directly on individual quantum effects. In these very same years when, in the development of his Theory of Relativity, Einstein laid down a new foundation for physical science, he explored with a most daring spirit the novel features of atomicity which pointed beyond the framework of classical physics.

Later, Bohr pointed out Einstein's own contributions to the statistical analysis of the quantum. In particular, Einstein

> formulated general statistical rules regarding the occurrence of radiative transitions between stationary states, assuming not only that, when the atom is exposed to a radiation field, absorption as well as emission processes will occur with a probability per unit time proportional to the intensity of the irradiation, but that even in the nascence of external disturbances spontaneous emission processes will take place with a rate corresponding to a certain *a priori* probability. Regarding the later point, Einstein emphasized the fundamental character of the statistical description in a most suggestive way.

Bohr went on to defend the quantum theory by outlining advances made in understanding and explaining the basic characteristics of the atom.

In the same volume, Einstein responded to criticisms of his stance on quantum theory. Einstein wrote:

> They are all firmly convinced that the riddle of the double nature of all corpuscles ... has in essence found its final conclusion in the statistical Quantum Theory. On the strength of the successes of this theory they consider it proved that a theoretically complete description of a system can, in essence, involve only statistical assertions concerning the measurable quantities of this system.

Expressing the concern of many scientists who found difficulties in accepting the statistical nature of the world, Einstein proposed that perhaps the problem is found in the incompleteness of the theory:

> I am, in fact, firmly convinced that the essentially statistical character of contemporary Quantum Theory is solely to be ascribed to the fact that this [theory] operates with an incomplete description of physical systems.
>
> ... Above all, however, the reader should be convinced that I fully recognize the very important progress which the statistical Quantum Theory has brought to theoretical physics.... The theory is until now the only one which unites the corpuscular and undulatory dual character of matter in a logically satisfactory fashion.... The formal relations which are given in this theory — i.e., its entire mathematical formulism — will probably have to be contained, in the form of logical inferences, in every useful future theory.
>
> ... What does not satisfy me in that theory, from the standpoint of principle, is its attitude towards that which appears to me to be the programmatic aim of all physics: the complete description of any (individual) real situation (as it supposedly exists irrespective of any act of observation or substantiation.) Whenever the positivistically inclined modern physicist hears such a formulation his reaction is that of a pitying smile. He says to himself: "There we have the naked formulation of a metaphysical prejudice, empty of content, a prejudice, moreover, the conquest of which constitutes the major epistemological achievement of physicists within the last quarter-century. Has any man ever perceived a 'real physical situation?' How is it possible that a reasonable person could today still believe that he can refute our essential knowledge and understanding by drawing up such a bloodless ghost?" Patience! The above laconic characterization was not meant to convince anyone; it was merely to indicate the point of view around which the following elementary considerations freely group themselves.

Einstein proceeded to make his arguments against the physical reality of the quantum theory. In answering the imaginary query of a physicist as to the possibility of finding a complete description of nature, independent of statistics, Einstein maintained:

> For me, however, the expectation that the adequate formulation of the universal laws involves the use of *all* conceptual elements which are necessary for a complete description, is more natural. It is furthermore not at all surprising that, by using an incomplete description, (in the main) only statistical statements can be obtained out of such a description. If it should be possible to move forward to a complete description, it is likely that the laws would represent relations among all the conceptual elements of this description which, *per se*, have nothing to do with statistics.

Albert Einstein and Niels Bohr were only two of a brilliant group of scientists who changed the fundamental tenets of physics in the first decades of the twentieth century. Contrary to Kelvin's dire prediction at the beginning of the century that there was little new to be learned about physics, the discoveries of Einstein, Bohr, and their contemporaries opened up a new world for generations of scientists to explore.

PRIMARY SOURCES

Bohr, Niels. "Discussions with Einstein on Epistemological Problems in Atomic Physics." *Albert Einstein: Philosopher-Scientist*, edited by Paul Arthur Schilpp. New York: Harper and Brothers, 1959. Reprinted in *The Tests of Time*, edited by Lisa M. Dolling, Arthur F. Gianelli, and Glenn N. Statile. Princeton: Princeton University Press, 2003.

_____. "On the Constitution of Atoms and Molecules," *Philosophical Magazine* 26, 1 (1913).

Einstein, Albert. "Einstein's Reply." *Albert Einstein: Philosopher-Scientist*, edited by Paul Arthur Schilpp. New York: Harper and Brothers, 1959. Reprinted in *The Tests of Time*, Lisa M. Dolling, Arthur F. Gianelli; and Glenn N. Statile. Princeton: Princeton University Press, 2003.

_____. *Relativity: The Special and General Theory*. Project Gutenberg. http://www.gutenberg.org/cache/epub/5001/pg5001.html.

Planck, Max. "The Genesis and Present State of Development of the Quantum Theory." Nobel Lecture, June 2, 1920. http://nobelprize.org/.

_____. *A Survey of Physical Theory*. New York: Dover, 1993.

Thomson, J.J. "Carriers of Negative Electricity." Nobel Lecture, December 11, 1906. http://nobelprize.org/.

OTHER SOURCES

Beller, Mara. *Quantum Dialogue: The Making of a Revolution*. Chicago: University of Chicago Press, 1999.

Bodanis, David. *Biography of the World's Most Famous Equation*. New York: Walker, 2000.

Brian, Denis. *Einstein: A Life*. New York: Wiley, 2006.

Calder, Nigel. *Einstein's Universe*. New York: Viking Press, 1979.

Galison, Peter. *Einstein's Clocks and Poincaré's Maps: Empires of Time*. New York: W.W. Norton, 2003.

Isaacson, Walter. *Einstein: His Life and Universe*. New York: Simon & Schuster, 2007.

Kaku, Michio. *Einstein's Cosmos: How Albert Einstein's Vision Transformed Our Understanding of Space and Time*. New York: W.W. Norton, 2004.

Kragh, Helge. *Quantum Generations: A History of Physics in the Twentieth Century*. Princeton: Princeton University Press, 1999.

Pais, Abraham. *Niels Bohr's Times: In Physics, Philosophy, and Polity*. New York: Oxford University Press, 1991.

_____. *"Subtle is the Lord": The Science and the Life of Albert Einstein*. Oxford and New York: Oxford University Press, 1982.

Pauli, Wolfgang, ed. *Niels Bohr and the Development of Physics; Essays Dedicated to Niels Bohr on the Occasion of His Seventieth Birthday*. New York: McGraw-Hill, 1955.

Sachs, Mendel. *Einstein Versus Bohr: The Continuing Controversies in Physics*. La Salle, IL: Open Court, 1988.

Whitaker, Andrew. *Einstein, Bohr, and the Quantum Dilemma*. Cambridge and New York: Cambridge University Press, 1996.

CHAPTER 13

From the Individual to the Collective: Oppenheimer, the Manhattan Project, and the Emergence of "Big Science"

In some sort of crude sense, which no vulgarity, no humor, no overstatement can quite extinguish, the physicists have known sin; and this is a knowledge which they cannot lose.—J. Robert Oppenheimer, *Time Magazine* (1948)

At the beginning of the twentieth century, scientists occupied essentially the same place as they had for thousands of years — on the fringe of society with no particular place to call their own. The research university with its emphasis on innovation in science and the arts was in its infancy, particularly so in the United States. With a few exceptions, industry had not yet found the need to employ scientists on a large-scale basis for research and development. And governments, again with a few exceptions, had yet to discover what sort of contributions scientists might make to defense or to a developing economy. By the end of the century, the situation was fundamentally and irreversibly changed. Scientists working at research universities, innovative industries, and governments across the globe were funded to an extent beyond the imagination of their predecessors just a few generations before. Their contributions to the economy of a nation, and especially to the defense of their homeland, were considered indispensable.

The world at war proved to be a major catalyst for this change in the very nature of the scientific enterprise. World War I stimulated advances in weapons (machine guns), as well as in modes of delivering weapons (tanks and airplanes). We also find the use of the first so-called weapon of mass destruction in World War I — poisonous gas. This new and horrible weapon, which killed indiscriminately and — with every subtle shift in wind direction — unpredictably, also represents one of the earliest contributions to war by scientists. First used in battle by the Germans, the Allies were soon employing chemists to develop their own gases to be used in retaliation.

While poison gas was a frightening new development in warfare, this contribution by scientists to military objectives proved to be only the beginning. World War II ushered in a new age of cooperation among political, military, and scientific leaders. Scientists for both the Allies and the Axis powers made dizzying advances in radar technology, ballistics, jet aircraft, and rocketry. Scientists in the United States began work on one of the earliest digital computers, the ENIAC (Electronic Numerical Integrator and Computer). Although predated by other primitive electronic computers, the ENIAC is considered the first general purpose electronic computer. Designed to calculate firing tables (ballistics), it also made some important calculations for the Manhattan Project, the code name for the project tasked to build the first atomic bombs.

The word computer was originally applied to human beings whose calculation skills made them indispensable to many scientific enterprises. During World War II, many of these computers were mathematically-trained women. In addition to work on the Manhattan Project, computers — both electronic and human — also played a major role in another important undertaking. The work of mathematicians, scientists, and linguists led to eventual success in breaking German codes. The Enigma was a machine designed by the German military to encrypt messages. The code produced by this machine was considered unbreakable by its designers; so, when the Allies eventually broke the unbreakable code, it presented a distinct advantage in the war effort. It has been argued that the ability to read German communications (and the fact that the Germans did not know that the Allies had broken the code) was the turning point in the war.

The scientists working on the aforementioned projects, along with many others, represented the brightest minds in the world. And they were, indeed, from all over the world, including Axis-occupied regions. In fact, as scientists, mathematicians, and intellectuals from all fields fled (or were forcibly exiled), the brain drain that hobbled the fascist regimes significantly added to the pool of talent in the United States and Britain. The scientists fleeing the Nazis and their allies represented a significant percentage of the university science professors, and Nobel Prize laureates, from countries such as Germany, Austria, Hungary, and Italy. Many of these immigrants contributed directly or indirectly to wartime projects in their adopted homelands.

Of all the advances made by scientists in support of the war efforts, the biggest, most famous, most expensive, and the one that changed the world most indelibly was the project to build the first atomic bomb. And although the idea for the bomb itself— as well as numerous contributions central to the success of the project — came from refugee scientists, the overall direction of the project fell to an American, J. Robert Oppenheimer. His leadership and guidance of hundreds of scientists, engineers, and technicians working for a common goal made Oppenheimer the central figure of the Manhattan Project.

Oppenheimer Before World War II

J. Robert Oppenheimer (1904–1967) was born in New York City to wealthy parents who, although Jewish, did not practice their religion. The precocious boy was educated at the finest schools New York had to offer, and entered Harvard in 1922 with interests in chemistry and physics. Oppenheimer graduated from Harvard, with honors, in 1925 and immediately set out for Europe to study with J.J. Thomson (awarded the Nobel Prize in 1906 for his discovery of the electron) at the renowned Cavendish Laboratory in Cambridge, England. Rather clumsy in the laboratory, it soon became apparent to both Oppenheimer and his teachers that theoretical, rather than experimental, physics was his future.

After a year of study at Cambridge, Oppenheimer moved to the University of Göttingen (Germany) to study with Max Born (who would win the Nobel Prize for physics in 1954). The University of Göttingen was at the epicenter of theoretical physics in the early twentieth century. In addition to Born, the best and brightest physics had to offer taught and studied at Göttingen while Oppenheimer was in residence. Werner Heisenberg (Nobel Prize, 1932), Wolfgang Pauli (Nobel Prize, 1945), Enrico Fermi (Nobel Prize, 1938), and Edward Teller (later to be called "father of the hydrogen bomb") were only a few of the world famous (or soon to be world famous) scientists with whom Oppenheimer learned and worked.

Receiving his Ph.D. from Göttingen in 1927 at the tender age of 22, Oppenheimer's world was full of career opportunities. He chose to return to the United States, eventually deciding to accept a joint appointment to both the California Institute of Technology (Cal Tech) and the University of California, Berkeley. Over the ensuing years, Oppenheimer's teaching and magnetic personality attracted many of the brightest America had to offer to California, establishing Oppenheimer as the father of American theoretical physics. Oppenheimer eventually married Katherine Peuning Harrison. Kitty, as she was known, had ties to the Communist Party through party membership and a previous husband who was killed in the Spanish Civil War. This relationship, along with other associations with left-wing groups and individuals, would lead to serious troubles for Oppenheimer many years later.

Throughout his career, Oppenheimer's research interests and publications varied widely over several subfields of physics. He did important work in subatomic particle physics and gravitational theory, where he became one of the first to predict the existence of black holes. Although his work never garnered him the accolades of some of his contemporaries in theoretical physics — in particular, Oppenheimer never received the Nobel Prize — his leadership role in the Manhattan Project indicates the sort of esteem which his fellow scientists held for him.

The Manhattan Project

World War II changed the very fabric of Robert Oppenheimer's life. Of course, the same may be said about millions of people all over the world, but Oppenheimer's role in developing the weapon that won the war defines the man and his legacy. The product of the Manhattan Project — the first atomic weapon — not only helped the United States swiftly end the war in the Pacific, it also instituted changes in military, political, and social contexts never before seen in history. Oppenheimer was at the center of these changes.

Even before President Roosevelt authorized the monumental task of creating an atomic weapon, Oppenheimer and other leading scientists, both American and refugees from war-torn Europe, were hard at work on ideas that proved critical to the bomb's development. Much of this work was centered at the Radiation Laboratory, or RadLab, operated by Ernest Lawrence at Berkeley. Lawrence (1901–1958) was born and educated in South Dakota, and awarded a Ph.D. in physics from Yale University. Unlike Oppenheimer (who worked with Lawrence at Berkeley), Lawrence's forte was experimental physics. He is best remembered for his invention of the cyclotron (or, as they are often called, atom smashers), machines designed to accelerate subatomic particles to high rates of speed. In 1939, Lawrence received the Nobel Prize in physics for his invention. The cyclotron proved vital to the exploration of subatomic particles and to the theoretical and practical development of the atomic bomb.

Before work on such a weapon could commence in earnest, however, it was necessary to inform political leaders that such a possibility existed.

> Some recent work by E. Fermi and L. Szilard, which has been communicated to me in manuscript, leads me to expect that the element uranium may be turned into a new and important source of energy in the immediate future.

These words penned in a letter to President Roosevelt in 1939 and signed by the most famous scientist in the world — Albert Einstein — marked the beginning of the nuclear age. The L. Szilard mentioned in the letter was Leo Szilard (1898–1964), a Hungarian refugee who was among the first to seriously consider the possibility of a nuclear chain reaction. Szilard proved the theoretical possibility of just such an event. In fact, it was Szilard himself who wrote the letter addressed to President Roosevelt, enlisting Einstein's help (and signature) to ensure that the letter would be read and taken seriously.

In the letter, Szilard and Einstein explained to Roosevelt that a nuclear chain reaction was possible, even imminent:

> In the course of the last four months it has been made probable through the work of Joliot in France as well as Fermi and Szilard in America that it may become possible to set up a nuclear chain reaction in a large mass of uranium, by which vast amounts of power and large quantities of new radium-like elements would be generated. Now it appears almost certain that this could be achieved in the immediate future.

This chain reaction, the letter continued, could potentially result in a new weapon the power of which has never before been imagined. Therefore, Szilard and Einstein urged, it would be wise to open lines of communication between the Roosevelt administration and the scientists working on chain reactions.

> This new phenomenon would also lead to the construction of bombs, and it is conceivable — though much less certain — that extremely powerful bombs of a new type may thus be constructed. A single bomb of this type, carried by boat and exploded in a port, might very well destroy the whole port together with some of the surrounding territory. However, such bombs might very well prove to be too heavy for transportation by air.... In view of the situation you may think it desirable to have more permanent contact maintained between the Administration and the group of physicists working on chain reactions in America.

The letter went on to outline a few suggestions for beginning research into the possibility of such a bomb. Szilard and Einstein closed with a statement meant to convey an extra sense of urgency by suggesting Germany was already beginning the process of developing a nuclear weapon — the implications being obvious to Roosevelt.

> I understand that Germany has actually stopped the sale of uranium from the Czechoslova-
> kian mines which she has taken over. That she should have taken such early action might
> perhaps be understood on the ground that the son of the German Under-Secretary of State,
> von Weizsäcker, is attached to the Kaiser-Wilhelm-Institut in Berlin where some of the
> American work on uranium is now being repeated.

Einstein and Szilard wrote two more letters to the president over the next eight months encouraging Roosevelt to begin serious work on the feasibility of such a weapon.

Fear of what might occur if Hitler obtained this terrible new weapon drove the scientists (along with the military leaders and politicians) to begin work on the atomic bomb. This fear was particularly acute in the refugee scientists who had observed first-hand the horrors of Nazi rule. Scientists from Germany, Hungary, Austria, and other countries falling under Nazi influence fled for Britain and the United States and in the process lent their talents to the war efforts of their adoptive countries. One Italian physicist, fleeing his homeland with his Jewish wife in the face of rising anti–Semitism, proved crucial to the first task at hand — creating a sustained and controlled nuclear chain reaction.

Enrico Fermi (1901–1954) was Italy's most important scientist. His experiments in which he fired neutrons at atomic nuclei and studied the resulting isotopes gained him world-wide acclaim by his fellow scientists. After traveling to Sweden in 1939 to receive a Nobel Prize for his discoveries, he and his wife immigrated to the United States where he continued his research at Columbia University in New York City. Soon after moving to the University of Chicago, Fermi was tapped to lead the team trying to produce a sustained nuclear reaction. In a laboratory built in a converted squash court underneath the stands of the University of Chicago football field, Fermi's team met with success. What exactly did Fermi and his team accomplish? Fermi made the analogy to a burning pile of trash:

> An atomic chain reaction may be compared to the burning of a rubbish pile from sponta-
> neous combustion. In such a fire, minute parts of the pile start to burn and in turn ignite
> other tiny fragments. When sufficient numbers of these fractional parts are heated to the kin-
> dling points, the entire heap bursts into flames.

To put into perspective the audacity of attempting such a feat, General Leslie Groves, director of the Manhattan Project, reminisced:

> At the laboratory in Chicago, we were seeking to split atoms, and in the process to transmute
> one element into another — that is, to change uranium into plutonium.... In effect, the scien-
> tists were reviving the classical, but always unsuccessful, search of the ancient alchemists for
> ways to convert base metals, such as lead, into gold; and the continuing, but theretofore
> unsuccessful, attempts of more modern chemists to change the character of the elements.
> The precedents of history were surely against us.

Fermi's atomic pile showed that a sustained reaction was possible and encouraged scientists that the development of an atomic bomb was on the horizon. Fermi continued as an integral part of the scientific team that eventually built and tested the first such weapon.

Much of the early research and planning for the first atomic weapons took place after the establishment of the Manhattan Engineering District of the Army Corps of Engineers in August of 1942, headquartered in New York City. With an increase in the sense of urgency — which resulted in increased funding — the Manhattan Engineering District expanded to various sites around the United States and was shortened to the Manhattan Project. The central site for research, development, and testing the bomb was moved to an isolated location in the high desert of New Mexico. Los Alamos became the birthplace of the atomic bomb. But other locations around the country played equally critical roles. Two of these locations, Oak Ridge, Tennessee, and Hanford, Washington, served as research laboratories and production facilities tasked with producing enough uranium and plutonium, respectively, to produce bombs. This task proved to be at least as complicated as the actual planning and building of the bomb itself. The quantities of pure uranium and pure plutonium needed for weapons had never before been produced, and no one was even sure what production methods, if any, would prove successful. Taken by themselves, both Oak Ridge and Hanford represent two of the largest scientific and engineering undertakings ever attempted by humankind. Together, along with the other facilities operated by the Manhattan Project, the United States embarked on a mission to complete the most complicated (and expensive) scientific project ever before attempted.

Oppenheimer: Leader of the Manhattan Project

Robert Oppenheimer's appointment as scientific director of the Manhattan Project was in many ways unexpected. Unlike many of his colleagues who also worked on the bomb, Oppenheimer's research had not resulted in the prestige of a Nobel Prize. Just as importantly, Oppenheimer had no real administrative experience that justified his selection to administer the largest scientific undertaking in history. Finally, Oppenheimer's ties to left wing and communist groups made his choice as director politically suspect. However, what Oppenheimer did have was the respect of the scientific community, and — most importantly — he made a favorable impression on General Leslie Groves, the Army engineer in charge of the Manhattan Project.

Oppenheimer's vision of the dynamics at Los Alamos — both scientific and social — was apparent in a letter addressed to James Conant. Conant, an accomplished chemist and president of Harvard for twenty years, played an integral role in organizing the Manhattan Project from his position as chairman of the National Defense Research Committee (NDRC). Conant, along with Vannevar Bush, the director of the Office of Scientific Research and Development (OSRD) and the first chairman of the NDRC, were critical liaisons between the U.S. government and the Manhattan Project scientists. In his letter to Conant, Oppenheimer expressed his desire to bring scientists to Los Alamos whom he knew and respected:

> There are, however, two reasons more substantial than prejudice why the limitation to men who are known to us is sound: 1) that the technical details of this work will in large part have to do with atomic physics so that any man whose experience had been in another field will necessarily be of more limited usefulness.... The second reason is that in a tight isolated group such as we are now planning, some warmth and trust in personal relations is an indis-

pensable prerequisite, and we are, of course, able to insure this only in the case of men whom we have known in the past.

As the research laboratory at Los Alamos was in its planning stages, recruiting the top-notch scientists to come to work on a top-secret project in the middle of the New Mexico high desert was Oppenheimer's primary mission. And, as any recruiter knows, a large part of Oppenheimer's job was soothing the fears and apprehensions of the spouses and families who would be sharing in the new life at Los Alamos. In a letter to physicist Hans Bethe, and his wife, Rose, Oppenheimer described the organization of the new city which would be created almost overnight:

> Now to Rose's question. There will be a sort of city manager who will be distinct from Colonel Harmon [the army officer who had current command of the construction and management of the new town]. Harmon was trying to find a man for this job and there is a man in New Mexico, a civilian, whom we could obtain if he is not successful. There will also be a city engineer and together they will take care of the problems outlined by you. We hope to persuade one of the teachers at the school to stay on to be our professional teacher. It is true that both Kay Manley and Elsie McMillan are professional school teachers and there will no doubt be others, but it seems to me unlikely that anyone with a very young child will be able to devote very much time to the community. There will be two hospitals, one in the town and one in the M.P. camp. The one in our town will be run by our doctor or doctors and she will be paid a regular salary so that she will be independent of our fees. I believe we shall have a group health plan of some kind but this is one of the questions that is tied up with how we handle subsistence.
>
> Room is being provided for a laundry; each house will have its washtub and we shall be able to send laundry to Santa Fe regularly. It may be necessary for us to provide the equipment for the group laundry since this is now frozen, but this is a point that is not yet settled.
>
> We plan to have two eating places. There will be a regular mess for unmarried people which will be, when we are running at full capacity, just large enough to take care of these. The Army will take care of the help for this and I do not know whether the personnel will be Army or civilian. We will also arrange to have a café where married people can eat out. This will probably be able to handle about twenty people at a time and will be a little fancy, and may be by appointment only. We are trying to persuade one of the natives to operate this and we have a good building for it.
>
> There will be a recreation officer who will make it his business to see that such things as libraries, pack trips, movies, and so on are taken care of, and he will no doubt welcome the help of as many of us as are willing. The bachelor apartments will be run by the Army and will be completely served. The store will be a so-called Post Exchange which is a combination of country store and mail order house. That is, there will be stocks on hand and the Exchange will be able to order for us what they do not carry. There will be a vet to inspect the meat and barbers and such like. There will also be a cantina where we can have beer and cokes and light lunches.

Robert Oppenheimer, one of the premiere physicists in the world, was happy to ensure that the needs of his fellow scientists and their families would be met.

As work on the bomb continued, it became ever more apparent that this project was one that, ultimately, would prove to be one of the most important scientific and military endeavors ever undertaken. Secrecy was always a priority; even as the scientists working at Los Alamos insisted on the right to free interchange of ideas, at least within the boundaries of the laboratory itself. The role that Oppenheimer played in the organization of the work was crucial, even indispensable. This was reflected in a letter from General Groves to Oppenheimer in the summer of 1943 in which the general asked Oppenheimer to take several precautions. Groves requested:

(a) You refrain from flying in airplanes of any description; the time saved is not worth the risk....

(b) You refrain from driving an automobile for any appreciable distance (above a few miles) and from being without suitable protection on any lonely road, such as the road from Los Alamos to Santa Fe. On such trips you should be accompanied by a competent, able bodied, armed guard....

(c) Your cars be driven with due regard to safety and that in driving about town a guard of some kind should be used, particularly during hours of darkness.

By 1943, General Groves was well aware that should anything happen to his scientific director, progress on the bomb would certainly be set back — an unacceptable scenario.

Looking back at Oppenheimer's appointment as scientific director, leading scientists working on the project expressed high praise for the way in which Oppenheimer performed such a stressful and critical job. Hans Bethe, the German immigrant and Nobel Prize laureate in physics, was an important contributor to the Manhattan Project. In his eulogy of Oppenheimer, Bethe remarked:

It was a marvelous choice. Los Alamos might have succeeded without him, but certainly only with much greater strain, less enthusiasm, and less speed. As it was, it was an unforgettable experience for all the members of the laboratory. There were other wartime laboratories of high achievement, like the Metallurgical Laboratory at Chicago, the Radiation Laboratory at M.I.T., and others, both here and abroad. But I have never observed in any of these other groups quite the spirit of belonging together, quite the urge to reminisce about the days of the laboratory, quite the feeling that this was really the great time of their lives.

Bethe quoted Victor Weisskopf, another important scientist who fled Nazism in Austria and became a leader at Los Alamos. Weisskopf remarked that Oppenheimer

did not direct from the head office. He was intellectually and even physically present at each decisive step. He was present in the laboratory or in the seminar rooms, when a new effect was measured, when a new idea was conceived. It was not that he contributed so many ideas or suggestions; he did so sometimes, but his main influence came from something else. It was his continuous and intense presence, which produced a sense of direct participation in all of us; it created that unique atmosphere of enthusiasm and challenge that pervaded the place throughout its time. He was everywhere at all times, and he worked incredibly long hours. Nevertheless, he still had time for some social life; in fact, the Oppenheimer house with his attractive wife was a social centre. He lived, as far as we could see, on his nervous energy. Always quite thin, he lost another twenty pounds and during a bout with measles reportedly got down to 104 lb., being six feet tall. It is remarkable that his health could stand this pace, because he was never physically strong. The one sport he loved was horse-back riding. But in the three years at Los Alamos there was time only for one overnight ride on the two horses his wife fed and groomed for their use. Before Los Alamos, on his ranch, he used to keep five horses for himself and his guests.

Bethe ended his eulogy of Oppenheimer with a succinct description of his superb mind:

Oppenheimer will be remembered by the world and by his country. He will leave a lasting memory in all the scientists who have worked with him, and in the many who have passed through his school and whose taste in physics was formed by him. His was a truly brilliant mind, best described by his long-time associate Charles Lauritsen: "This man was unbelievable. He always gave you the answer before you had time to formulate the question."

In retrospect, General Groves' selection of J. Robert Oppenheimer was as important as any single scientific discovery or breakthrough for the eventual success of the Manhattan Project.

The Decision to Use the Bomb

In May of 1944, a little more than one year before the scientists of the Manhattan Project successfully tested the world's first nuclear weapon, Niels Bohr — perhaps the second most famous scientist in the world after Einstein — addressed a letter to Winston Churchill offering his evaluation of the project and its expected outcome. Bohr's letter to Churchill outlined the immensity of the project and the progress being made on the atomic bomb:

> In fact, what until a few years ago might be considered as a fantastic dream is at present being realized within great laboratories and huge production plants secretly erected in some of the most solitary regions of the United States. There a larger group of physicists than ever before collected for a single purpose, working hand in hand with a whole army of engineers and technicians, are preparing new materials capable of an immense energy release, and are developing ingenious devices for the most effective use of these materials.

Bohr then broached the subject that was more and more on the minds of the physicists who were beginning to contemplate the magnitude of what they were doing:

> The whole undertaking constitutes, indeed, a far deeper interference with the natural course of events than anything ever before attempted, and it must be realized that the success of the endeavors has created a quite new situation as regards human resources. The revolution in industrial development which may result in coming years cannot at present be surveyed, but the fact of immediate preponderance is, that a weapon of devastating power far beyond any previous possibilities and imagination will soon become available.

Bohr, confident that the work done in the Manhattan Project would yield a "revolution in industrial development," followed this immediately in his letter by introducing to Churchill what was on the minds of many of the scientists involved with the bomb — how will this new weapon be controlled so as not to lead to the destruction of mankind?

After the end of the war, Niels Bohr became one of the leading scientists advocating for international control of nuclear energy and nuclear weapons. He planted the seeds for such an advocacy by asking Churchill to consider how questions of control and proliferation would be dealt with, and whether the scientists who created the technology should be included in the political process.

> The lead in the efforts to master such mighty forces of nature, hitherto beyond human reach, which by good fortune has been achieved by the two great free nations, entails the greatest promise for the future. The responsibility for handling the situation rests, of course, with the statesmen alone. The scientists who are brought into confidence can only offer the statesmen all such information about the technical matters as may be of importance for their deci-

sions.... These circumstances obviously have an important bearing on the question of an eventual competition about the formidable weapon, and on the problem of establishing an effective control, and might therefore perhaps influence the judgment of the statesmen as to how the present favorable situation can best be turned to lasting advantage for the cause of freedom and world security.

Bohr, of course, was not the only scientist concerned about the future of nuclear weapons, even before the first such weapon was completed. In a letter from a committee of scientists working at the Metallurgical Laboratory, or Met Lab, in Chicago (an important component of the Manhattan Project), the chairman of the committee, James Franck, began with a simple, yet ominous, prediction:

> The development of nuclear power not only constitutes an important addition to the technological and military power of the United States, but also creates grave political and economic problems for the future of this country.

The letter went on to accurately predict an arms race when other countries inevitably develop their own nuclear weapons:

> Nuclear bombs cannot possibly remain a "secret weapon" at the exclusive disposal of this country, for more than a few years. The scientific facts on which their construction is based are well known to scientists of other countries. Unless an effective international control of nuclear explosives is instituted, a race of nuclear armaments is certain to ensue following the first revelation of our possession of nuclear weapons to the world. Within ten years other countries may have nuclear bombs, each of which, weighing less than a ton, could destroy an urban area of more than ten square miles.

The letter recommended that the Scientific Panel of the Interim Committee (a committee of government officials and scientists appointed by President Truman to discuss and advise him on nuclear policies) ask for a demonstration of the bomb and to make public the possession of such a weapon.

In the Scientific Panel's recommendation to the Interim Committee, signed by Oppenheimer on the behalf of the entire panel, they acknowledge that there was not a consensus among scientists as to whether or not the new weapon should be used. However, the panel agreed with those who "emphasize the opportunity of saving American lives by immediate military use." Furthermore, the panel could "see no acceptable alternative to direct military use." "No acceptable alternative" was the conclusion of the Scientific Panel. But the decision to use the bomb did not seem that simple to all who were involved in 1945.

Many factors were considered before the final decision to use the bomb on Japanese targets was finalized. There were certainly a multitude of moral factors involved. This terrible new weapon did not differentiate between enemy combatants and innocent civilians. Even if military targets were chosen, the wide-ranging destructive properties of an atomic weapon guaranteed that countless civilians would be killed and injured. It was argued, however, that the widespread firebombing of Tokyo had already resulted in thousands of civilian fatalities; so, the morality of killing innocents was already breached.

Consideration was given to the idea that Japan should be warned before the bomb was used. Perhaps a demonstration of the weapon's capabilities would shock the Japanese leaders into surrendering. What if, however, the test did not work? Even if it did work, there was only enough material on hand to make three bombs. Therefore, if the scare tactic failed to induce surrender it meant one less weapon left for military use. Furthermore, it was argued that the Japanese had provided no warning before the attack on Pearl Harbor and thus deserved no warning of their own.

If it was decided that the bomb would be used, one political question was whether to consult (or even inform) the other Allies of its impending deployment. In particular, what were the strategic ramifications of notifying Russia? The developing political situation made it obvious that Russia and the United States would not remain allies after the war — in fact they would most certainly become rivals, if not enemies. In the end, this political situation provided one of the major motivations for dropping the bomb on Japan without delay. Because Russia had not yet entered the war against Japan, a quick end to the conflict would result in the United States dictating the peace without sharing power with Russia as they were in Europe.

In addition to the Russian question, other politics entered into the final decision to use the bomb. The Manhattan Project had cost about two billion dollars — an astronomical sum in the 1940s. If the bomb did end the war with Japan, it could save hundreds of thousands of lives, both American and Japanese, and justify the immense sum spent on developing the new weapon. In the end, most historians agree that President Harry Truman (President Roosevelt died before the final test of the atomic bomb) had little choice but to authorize the use of the first nuclear weapons on Japan.

When the first atomic bomb was tested at the test site code-named Trinity near Alamogordo, New Mexico, on July 16, 1945, it was immediately apparent that this group of scientists, engineers, and technicians had changed the world. Oppenheimer later related that upon witnessing the awesome display of power he immediately thought of a quote from the sacred Hindu text, the Bhagavad-Gita: "I am become death, the shatterer of worlds." Only a few weeks later, on August 6, the world of thousands living in and around Hiroshima, Japan, was shattered by the first atomic bomb used against an enemy target. "Little Boy," a uranium bomb, immediately killed up to 100,000 people, and ultimately resulted in the deaths of tens of thousands more. Three days later, the plutonium bomb "Fat Man" was detonated over Nagasaki, resulting in another 50,000 to 80,000 deaths. On September 2, under the threat of more attacks, Japan surrendered.

In the days, weeks, and years after Hiroshima and Nagasaki, most Americans agreed that the terrible destruction and loss of life was justified because it brought a speedy end to the war. Although it was argued that Japan was on the verge of surrender before the atomic bombs were unleashed and that the unthinkable devastation was morally indefensible, most of the scientists and politicians involved in the building of the bombs and the decision to use them held fast to their belief that it was a necessity of war. That is not to say that these leaders harbored no doubts or misgivings over their roles in ushering in an uncertain new age. In an interview with *Time Magazine* in 1948, Oppenheimer himself declared:

> In some sort of crude sense, which no vulgarity, no humor, no overstatement can quite
> extinguish, the physicists have known sin; and this is a knowledge which they cannot lose.

Oppenheimer's profound contributions to the war effort, unfortunately, would prove to be too little to save him from the Red Scare that swept the country in the years to come. After the war, Congress formed the Atomic Energy Commission (AEC), a civilian entity tasked with control over the peacetime development of nuclear energy. Oppenheimer was named chair of the General Advisory Committee (GAC), a committee formed to give technical and scientific advice to the AEC. Oppenheimer's many ties, both direct and indirect, to the Communist Party and to students, friends, and relatives (including his own wife, Kitty) who had been active in the Communist Party, was a continuing cause for concern to the American authorities even as the physicist was leading the Manhattan Project. His importance to the successful outcome of the project, however, seemed to always override the concerns of the military and the FBI. As chair of the GAC, however, Oppenheimer advised against the development of the hydrogen bomb — a fusion type weapon often referred to as the Super — and his political and scientific enemies grasped the opportunity to accuse Oppenheimer of disloyalty to the United States. After months of hearings, Oppenheimer's security clearance was suspended and his work with the United States government came to an end in 1954. Oppenheimer continued his work as a physicist and public lecturer until his death from throat cancer in 1967.

Post-War Science

The Manhattan Project and the other major science initiatives that played such important roles in determining the outcome of World War II changed forever the landscape of science in America, and indeed the world. With a price tag of $2 billion dollars, the Manhattan Project dwarfed the expenditures of any other scientific or military project before. No longer could science and scientists be left to their own devices — their work was now integral to national security. This realization began to emerge even before the end of the war, when President Roosevelt asked Vannevar Bush to make recommendations concerning the continuing cooperation between the government and the scientific establishment. In a report addressed to President Roosevelt, dated July 1945 (but submitted to President Truman, as Roosevelt had died a few months earlier), Bush outlined a plan to make permanent the new-found support for science by the federal government. This report, titled *Science, the Endless Frontier*, responded to a set of direct queries and suggestions made by Roosevelt:

> What can be done, consistent with military security, and with the prior approval of the military authorities, to make known to the world as soon as possible the contributions which have been made during our war effort to scientific knowledge?
> With particular reference to the war of science against disease, what can be done now to organize a program for continuing in the future the work which has been done in medicine and related sciences?
> What can the Government do now and in the future to aid research activities by public and private organizations?
> Can an effective program be proposed for discovering and developing scientific talent in American youth so that the continuing future of scientific research in this country may be assured on a level comparable to what has been done during the war?
> New frontiers of the mind are before us, and if they are pioneered with the same vision,

boldness, and drive with which we have waged this war we can create a fuller and more fruitful employment and a fuller and more fruitful life.

The pioneer spirit is still vigorous within this nation. Science offers a largely unexplored
hinterland for the pioneer who has the tools for his task. The rewards of such exploration
both for the Nation and the individual are great. Scientific progress is one essential key to
our security as a nation, to our better health, to more jobs, to a higher standard of living,
and to our cultural progress.

Bush's report included the need to increase the "scientific capital" in America by
increased attention to scientific education:

First, we must have plenty of men and women trained in science....
Second, we must strengthen the centers of basic research which are principally the colleges,
universities, and research institutes ... most research in industry and Government involves
application of existing scientific knowledge to practical problems. It is only the colleges, universities, and a few research institutes that devote most of their research to expanding the
frontiers of knowledge.

Bush continued by noting the respective roles of the government and the scientist if this
partnership were to continue:

The most important ways in which the Government can promote industrial research are to
increase the flow of new scientific knowledge through support of basic research, and to aid in
the development of scientific talent.

The responsibility for the creation of new scientific knowledge — and for most of its application — rests on that small body of men and women who understand the fundamental laws
of nature and are skilled in the techniques of scientific research.

And finally, Bush called for a program of action that would ensure that this vision of post-
war science would flourish:

The Government should accept new responsibilities for promoting the flow of new scientific
knowledge and the development of scientific talent in our youth. These responsibilities are
the proper concern of the Government, for they vitally affect our health, our jobs, and our
national security. It is in keeping also with basic United States policy that the Government
should foster the opening of new frontiers and this is the modern way to do it. For many
years the Government has wisely supported research in the agricultural colleges and the
benefits have been great. The time has come when such support should be extended to other
fields.

Therefore I recommend that a new agency for these purposed be established. Such an
agency should be composed of persons of broad interest and experience, having an understanding of the peculiarities of scientific research and scientific education. It should have the
stability of funds so that long-range programs may be undertaken. It should recognize that
freedom of inquiry must be preserved and should leave internal control of policy, personnel,
and the method and scope of research to the institutions in which it is carried out. It should
be fully responsible to the President and through him to the Congress for its program.

Although not fully realized for some time to come, with the establishment of the National
Science Foundation in 1950, Bush's dream for a stable agency supported by federal funds
and dedicated to basic and applied research came to fruition.

Vannevar Bush's vision of post-war science recognized that a new era was at hand
for science and for scientists. Over the centuries, science had been predominantly a solitary

pursuit. The images of Galileo alone with his telescope, Newton alone with his calculations, or Darwin alone with his thoughts, represented the ideal of *doing* science. Although exceptions existed (Brahe and his fully-staffed observatory, to name one), the image persisted — scientists were lonely visionaries pursuing their individual quests to understand nature.

The twentieth century, especially the latter half of the century, found this common conception of science (and scientists) inexorably changed. Scientists accustomed to wartime funding sought to continue their research with the aid of government funds. On their part, the politicians awoke to the military and economic benefits from scientific research and willingly committed ever increasing funds for this research. The era of Big Science was born. Big Science, of which the Manhattan Project is one of the first and best known examples, involves large numbers of scientist and technicians working towards a single goal — or at least a very narrow set of goals. And, of course, these Big Science projects meant big money; money that could only be provided by governments, or possibly large industrial concerns.

The Manhattan Project set the tone for other Big Science projects. Government agencies (and the scientists who benefited from their funding) became convinced that this model of science could answer questions and solve many of the problems facing society. Medical research targeted the cure or the eradication of various diseases from polio to cancer. Space projects from the Mercury program to the Apollo program directed large sums of resources, both human and financial, towards a single-minded goal. The Human Genome Project employed armies of scientists and technicians, as well as sophisticated and expensive instruments and laboratories, to map the human genome. Each of these projects (and many others) pointed the way towards a new way of understanding the roles of the scientist and the government for scientific research. Twentieth century science was as different from the science of Galileo as Galileo's work was from that of Aristotle.

PRIMARY SOURCES

Bethe, H. A. "J. Robert Oppenheimer. 1904–1967." *Biographical Memoirs of Fellows of the Royal Society*, Vol. 14 (Nov. 1968), pp. 391–416.

Kelly, Cynthia C. *The Manhattan Project: The Birth of the Atomic Bomb in the Words of Its Creators, Eyewitnesses, and Historians.* New York: Black Dog & Levanthal, 2007.

Smith, Alice Kimball, and Charles Weiner, eds. *Robert Oppenheimer: Letters and Recollections.* Stanford: Stanford University Press, 1980.

OTHER SOURCES

Alperovitz, Gar. *The Decision to Use the Atomic Bomb.* New York: Alfred A. Knopf, 1995.

Badash, Lawrence. *Scientists and the Development of Nuclear Weapons.* Amherst, NY: Prometheus Books, 1995.

Bernstein, Jeremy. *Oppenheimer: Portrait of an Enigma.* Chicago: Ivan R. Dee, 2004.

Bird, Kai, and Martin J. Sherwin. *American Prometheus: The Triumph and Tragedy of J. Robert Oppenheimer.* New York: Alfred A. Knopf, 2005.

Cassidy, David C. *J. Robert Oppenheimer and the American Century.* New York: Pi Press, 2005.

Conant, Jennet. *109 East Palace: Robert Oppenheimer and the Secret City of Los Alamos.* New York: Simon & Schuster, 2005.

Herken, Gregg. *Brotherhood of the Bomb: The Tangled Lives and Loyalties of Robert Oppenheimer, Ernest Lawrence, and Edward Teller.* New York: Henry Holt, 2002.

Hershberg, James G. *James B. Conant: Harvard to Hiroshima and the Making of the Nuclear Age.* New York: Knopf, 1993.

Hughes, Jeff. *The Manhattan Project: Big Science and the Atom Bomb.* New York: Columbia University Press, 2002.

McMillan, Priscilla Johnson. *The Ruin of J. Robert Oppenheimer and the Birth of the Modern Arms Race.* New York: Viking, 2005.

Oppenheimer, J. Robert, and I. I. Rabi. *Oppenheimer.* New York: Scribner, 1969.

Pais, Abraham, and Robert P. Crease. *J. Robert Oppenheimer: A Life.* Oxford and New York: Oxford University Press, 2006.

Rhodes, Richard. *The Making of the Atomic Bomb.* New York: Simon & Schuster, 1986.

Schweber, S. S. *In the Shadow of the Bomb: Bethe, Oppenheimer, and the Moral Responsibility of the Scientist.* Princeton: Princeton University Press, 2000.

CHAPTER 14

Genetics, Germ Theory, and DNA: The Work of Mendel, Pasteur, Watson and Crick

The striking regularity with which the same hybrid forms always reappeared whenever fertilization took place between the same species induced further experiments to be undertaken, the object of which was to follow up the developments of the hybrids in their progeny. — Gregor Mendel, "Lecture at the Natural History Society of Brünn" (1865)

If it is a terrifying thought that life is at the mercy of the multiplication of these minute bodies, it is a consoling hope that Science will not always remain powerless before such enemies, since for example at the very beginning of the study we find that simple exposure to air is sufficient at times to destroy them. — Louis Pasteur, "The Germ Theory and its Applications to Medicine and Surgery" (1878)

We wish to put forward a radically different structure for the salt of deoxyribose nucleic acid. This structure has two helical chains each coiled round the same axis. — James Watson and Francis Crick, "A Structure for Deoxyribose Nucleic Acid" (1953)

Two mysteries of the fundamental nature of life eluded natural philosophers for millennia. The first question was "What is the root cause of disease?" and the second was "How are traits passed from generation to generation?" On the first question, physicians and scientists in the nineteenth century began to debate the causes of disease. Throughout much of history the four humour theory of Hippocrates dominated western thought. Hippocrates identified these four humors (black bile, yellow bile, phlegm, and blood) and claimed that disease and sickness occurred when an imbalance of the humors occurred in the body. Thus, treatments for diseases included bloodletting, expectorants, laxatives, diuretics, and similar applications to expel the offending humour. Although there were a few who postulated other causes for disease — including small, unseen particles entering the body — it was not until the work of Louis Pasteur, Joseph Lister, Robert

Koch, and other nineteenth century scientists and physicians that the germ theory of disease was firmly established.

On the question of inheritance, a theory first suggested by ancient Greek philosophers — including Aristotle — came to be known as the inheritance of acquired characteristics. This theory maintained that when a living organism acquired a new trait during its lifetime that trait would be inherited by its offspring. Although Aristotle did not believe these new traits could go so far as to actually change the species, in the nineteenth century the theory was expanded by Lamarck and laid the groundwork for evolution and inspired the work of Darwin on natural selection. Beginning with an obscure Austrian monk by the name of Gregor Mendel — considered the father of genetics — and continuing through the discovery of the structure of the DNA molecule by James Watson and Francis Crick, biologists began to unravel the mystery of the mechanism of inheritance.

Gregor Mendel, the Father of Genetics

> Experience of artificial fertilization, such as is effected with ornamental plants in order to obtain new variations in color, has led to the experiments which will here be discussed. The striking regularity with which the same hybrid forms always reappeared whenever fertilization took place between the same species induced further experiments to be undertaken, the object of which was to follow up the developments of the hybrids in their progeny.

With these words, a relatively unknown monk by the name of Gregor Mendel opened his lecture at the Natural History Society of Brünn (now in the Czech Republic) in 1865. Nobody present at the meeting — or for that matter anyone who read Mendel's published manuscript — understood the significance of the results. In fact, it would be several decades before the work of Mendel was discovered and his place as the father of genetics was secured.

Born Johann Mendel (1822–1884) to an ethnically German family in a part of the Austrian empire that is now the Czech Republic, Mendel's childhood was spent on the family farm. Although not affluent by any means, young Mendel's family sacrificed greatly to pay for his education. He took the orders of the church and became an Augustine monk, where he took the name Gregor and began teaching. When his talents were recognized by his superiors, the young monk was sent to continue his studies at the University of Vienna. Mendel returned to his abbey at Brünn where he taught for fifteen years before becoming abbot of the monastery. During his time as a teacher at Brünn, Mendel performed the experiments that would someday make his name famous. Upon becoming abbot, however, Mendel's scientific work effectively ended as the business of the abbey occupied most of his time.

At first interested in breeding and studying mice, Mendel soon turned his attention to pea plants, which he cultivated in the monastery gardens. Mendel understood that he was embarking on a long and tedious study:

> It requires indeed some courage to undertake a labor of such far-reaching extent; this appears, however, to be the only right way by which we can finally reach the solution of a question the importance of which cannot be overestimated in connection with the history of the evolution of organic forms.

Mendel continued by explaining how his experiments were conducted and why he chose the pea plant for his studies:

> The paper now presented records the results of such a detailed experiment. This experiment was practically confined to a small plant group, and is now, after eight years' pursuit, concluded in all essentials. Whether the plan upon which the separate experiments were conducted and carried out was the best suited to attain the desired end is left to the friendly decision of the reader.
>
> At the very outset special attention was devoted to the *Leguminosae* on account of their peculiar floral structure. Experiments which were made with several members of this family led to the result that the genus *Pisum* was found to possess the necessary qualifications.
>
> Some thoroughly distinct forms of this genus possess characters which are constant, and easily and certainly recognizable, and when their hybrids are mutually crossed they yield perfectly fertile progeny. Furthermore, a disturbance through foreign pollen cannot easily occur, since the fertilizing organs are closely packed inside the keel and the anthers burst within the bud, so that the stigma becomes covered with pollen even before the flower opens. This circumstance is especially important. As additional advantages worth mentioning, there may be cited the easy culture of these plants in the open ground and in pots, and also their relatively short period of growth. Artificial fertilization is certainly a somewhat elaborate process, but nearly always succeeds. For this purpose the bud is opened before it is perfectly developed, the keel is removed, and each stamen carefully extracted by means of forceps, after which the stigma can at once be dusted over with the foreign pollen.
>
> In all, 34 more or less distinct varieties of Peas were obtained from several seedsmen and subjected to a two year's trial. In the case of one variety there were noticed, among a larger number of plants all alike, a few forms which were markedly different. These, however, did not vary in the following year, and agreed entirely with another variety obtained from the same seedsman; the seeds were therefore doubtless merely accidentally mixed. All the other varieties yielded perfectly constant and similar offspring; at any rate, no essential difference was observed during two trial years. For fertilization 22 of these were selected and cultivated during the whole period of the experiments. They remained constant without any exception.

Mendel was convinced that heredity was not haphazard and subject to chance, but rather followed certain laws of nature. His experiments were designed to discover these laws.

After eight years of experimentation and study, which included almost 30,000 pea plants, he developed what are known today as Mendel's laws of inheritance. The first law, called the segregation law, says that any gamete (a cell for sexual reproduction) contains only one characteristic. These characteristics, according to Mendel, were heredity units he called factors, later known as genes. Although these factors occur in pairs in most cells, in gametes they occur individually. So, for instance, when Mendel crossed plants whose pods where yellow with green pod plants, all of the resulting plants were green. However, when Mendel caused these first generation green pod plants to self-fertilize, some of the offspring were green while others were yellow. The fascinating fact, however, was that after thousands of experiments, Mendel found the ratio of green to yellow pods was almost exactly 3 to 1. His conclusion was that green was a dominant characteristic that hid the recessive characteristic of yellow in the first generation of offspring. However, the recessive characteristics remerged — in a very precise, mathematical way — in the second generation. Mendel's explanation of his discovery of this law is the root of what became Mendelian genetics.

Experiments which in previous years were made with ornamental plants have already afforded evidence that the hybrids, as a rule, are not exactly intermediate between the parental species. With some of the more striking characters, those, for instance, which relate to the form and size of the leaves, the pubescence of the several parts, etc., the intermediate, indeed, is nearly always to be seen; in other cases, however, one of the two parental characters is so preponderant that it is difficult, or quite impossible, to detect the other in the hybrid.

This is precisely the case with the Pea hybrids. In the case of each of the 7 crosses the hybrid-character resembles that of one of the parental forms so closely that the other either escapes observation completely or cannot be detected with certainty. This circumstance is of great importance in the determination and classification of the forms under which the offspring of the hybrids appear. Henceforth in this paper those characters which are transmitted entire, or almost unchanged in the hybridization, and therefore in themselves constitute the characters of the hybrid, are termed the *dominant*, and those which become latent in the process *recessive*. The expression "recessive" has been chosen because the characters thereby designated withdraw or entirely disappear in the hybrids, but nevertheless reappear unchanged in their progeny, as will be demonstrated later on.

Those forms which in the first generation exhibit the recessive character do not further vary in the second generation as regards this character; they remain constant in their offspring.

It is otherwise with those which possess the dominant character in the first generation. Of these *two*-thirds yield offspring which display the dominant and recessive characters in the proportion of 3:1, and thereby show exactly the same ratio as the hybrid forms, while only *one*-third remains with the dominant character constant.

The ratio of 3:1, in accordance with which the distribution of the dominant and recessive characters results in the first generation, resolves itself therefore *in all experiments into the ratio of 2:1:1,* if the dominant character be differentiated according to its significance as a hybrid-character or as a parental one. Since the members of the first generation spring directly from the seed of the hybrids, *it is now clear that the hybrids form seeds having one or other of the two differentiating characters, and of these one-half develop again the hybrid form, while the other half yield plants which remain constant and receive the dominant or the recessive characters in equal numbers.*

Mendel also discovered what is now known as his second law, the law of independent assortment. Mendel found that when two characteristics are inherited by an offspring, the individual characteristics are independent of each other. For example, the factor that determined the *color* of a pea plant was inherited independently from the factor that determined the *size* of the plant.

After his work with pea plants, Mendel later turned his attention to bees, crossbreeding and studying their genetic characteristics. However, very little is left of these studies, as Mendel never published his findings. The many duties of his new position as abbot of the monastery, including ongoing battles with government officials over what Mendel considered unfair taxes placed upon religious institutions, meant that Mendel had little time for science in the later years of his life.

Mendel published his study of the genetics of the pea plant in the *Transactions of the Brünn Natural History Society*. It was not a widely circulated journal, and few of Mendel's contemporaries in natural history read of his research and results. And, unfortunately, it seems that the few who did read Mendel's work did not grasp its significance. Mendel died in relative obscurity, the title "father of genetics" still many years in the future.

It was not until several decades later that three scientists (Hugo de Vries, Erich von Tschermak-Seysenegg and Carl Correns) independently reached conclusions similar to Mendel's and, in the process, discovered Mendel's work. De Vries, professor of botany at the University of Amsterdam, gave Mendel the posthumous credit he should have received decades before:

> This simple law for the constitution of the second generation of varietal hybrids with a single differentiating mark in their parents is called the law of Mendel. Mendel published it in 1865, but his paper remained nearly unknown to scientific hybridists. It is only of late years that it has assumed a high place in scientific literature, and attained the first rank as an investigation on fundamental questions of heredity. Read in the light of modern ideas on unit characters it is now one of the most important works on heredity and has already widespread and abiding influence on the philosophy of hybridism in general.
>
> But from its very nature and from the choice of the material made by Mendel, it is restricted to balanced or varietal crosses. It assumes pairs of characters and calls the active unit of the pair dominant, and the latent recessive, without further investigations of the question of latency. It was worked out by Mendel for a large group of varieties of peas, but it holds good, with only apparent exceptions, for a wide range of cases of crosses of varietal characters. Recently many instances have been tested, and even in many cases third and later generations have been counted, and whenever the evidence was complete enough to be trusted, Mendel's prophecy has been found to be right.

The work of de Vries and Correns, along with a new generation of scientists in the early twentieth century (most notably Thomas Hunt Morgan, Theodosius Dobzhansky, and Ernst Mayr), led to the modern synthesis, an account of biological evolution that combined Darwinian natural selection with Mendelian genetics.

The second half of the nineteenth century witnessed the birth of two of the most important theories in the life sciences — natural selection and Mendelian genetics. At about the same time, several scientists and physicians were leading the efforts to prove a critical theory — the germ theory of disease. Although many contributions were vital to the eventual acceptance of this controversial theory, perhaps no one name was more important than Louis Pasteur.

Louis Pasteur

Born into a poor French farming family, Louis Pasteur (1822–1895) was fortunate to receive a first-class education thanks in part to a teacher who recognized his talents and encouraged young Pasteur to seek higher education. The young man from humble beginnings eventually received a doctorate in chemistry from the Ecole Normale Superieure. Pasteur married and had five children. Three of his children died of typhoid, motivating Pasteur in his life-long search for the cause and prevention of deadly diseases.

At the age of 32, Pasteur was appointed dean of sciences at the University of Lille. As luck would have it, this region of France was a leading producer of beer. When a local brewery began experiencing severe problems with spoiling beer, they approached Pasteur for help. After examining samples of the brew under his microscope, Pasteur became convinced that the beer was souring because of the microscopic creatures he observed. He realized that these microbes could, and did, adversely affect the quality of the beer. Later,

Pasteur connected the same cause to the souring of wine and milk. He found that gentle heating (just enough to slow the growth of the microbes but not spoil the liquid itself) produced the desired effects — pasteurization was born. At the same time, the young scientist began to wonder if these same microscopic organisms might play a role in diseases contracted by animals and humans.

In 1857 Pasteur accepted a position at his alma mater, the Ecole Normale Superieure. In a series of ingenious experiments, Pasteur showed once and for all that the theory of spontaneous generation was false. Since the ancient Greeks, people had believed that life appeared from nowhere in the presence of such things as rotting meat. And, it wasn't only small creatures such as maggots in rotting meat that were thought to appear spontaneously. Medieval beliefs held that mice were produced from rotting grains, frogs from mud, and so forth. Although the seventeenth century Italian physician Francesco Redi performed experiments that showed maggots only appeared in rotting meat when left uncovered and accessible to flies, two centuries later many scientists continued to hold on to the idea that microscopic life could spontaneously appear in non-living materials. Pasteur's experiments with glassware specially designed to keep microbes in the air from coming into contact with the liquids finally put to rest the theory of spontaneous generation. Spurred on by his success, Pasteur later extended his works to microbes invading living organisms, and eventually showed that these microbes were responsible for diseases in animals and humans.

The Germ Theory of Disease

The cause of disease had forever baffled humankind. Theories abound concerning what made us sick. Was it an imbalance of the four humors? Was it bad airs emanating from sewage or swamps and bogs? Was it a misalignment of the stars or other astrological causes? Or was disease and sickness a punishment for sin? The discovery that it was none of these, but rather microscopic particles hidden away from our view marks the beginning of modern scientific medicine.

Although it was the later part of the nineteenth century before the germ theory of disease was seriously studied and slowly accepted by the medical community, germs as the cause of disease had been proposed before. In the sixteenth century, for instance, the Italian physician Giralamo Francastoro postulated that epidemic diseases were caused by invisible seeds that multiplied in the body. Even the great Swedish naturalist Carl Linnaeus proposed that small particles were the root cause of disease. These ideas, however, had little influence as no experimental or observational evidence was provided to substantiate the theories.

A fundamental advance towards an understanding of the causes of disease came with the invention of the microscope. Developed by Dutch eyeglass makers late in the sixteenth century, in the hands of scientists such as Robert Hooke and Anton van Leeuwenhoek the microscope became as indispensable for studying the tiny and close by as the telescope was to the study of the large and far away. In particular, Hooke's observation of the structure of cells and van Leeuwenhoek's discovery of bacteria laid a foundation for physicians and physiologists to explore the impact of microscopic creatures on the human health.

The invention of the microscope opened a flurry of observations by men like Hooke and van Leeuwenhoek. These scientists and others like them rushed to observe everything in their surroundings, from a drop of water to everyday objects around the laboratory. In the foremost book on the use of microscopes (*Micrographia*), Hooke wrote:

We will begin these our Inquiries therefore with the Observations of Bodies of the most *simple nature* first, and so gradually proceed to those of a more *compounded* one. In prosecution of which method, we shall begin with a *Physical point*; of which kind the *Point of a Needle* is commonly reckon'd for one; and is indeed, for the most part, made so sharp, that the naked eye cannot distinguish any parts of it: It very easily pierces, and makes its way through all kind of bodies softer then it self: But if view'd with a very good *Microscope*, we may find that the *top* of a Needle (though as to the sense very *sharp*) appears a *broad, blunt,* and very *irregular* end; not resembling a Cone, as is imagin'd, but only a piece of a tapering body, with a great part of the top remov'd, or deficient. The Points of Pins are yet more blunt, and the Points of the most curious Mathematical Instruments do very seldom arrive at so great a sharpness; how much therefore can be built upon demonstrations made only by the productions of the Ruler and Compasses, he will be better able to consider that shall but view those *points* and *lines* with a *Microscope.*

Now though this point be commonly accounted the sharpest (whence when we would express the sharpness of a point the most *superlatively,* we say, As sharp as a Needle) yet the *Microscope* can afford us hundreds of Instances of Points many thousand times sharper: such as those of the *hairs,* and *bristles,* and *claws* of multitudes of *Insects;* the *thorns,* or *crooks,* or *hairs* of *leaves,* and other small vegetables; nay, the ends of the *stiriæ* or small *parallelipipeds* of *Amianthus,* and *alumen plumosum;* of many of which, though the Points are so sharp as not to be visible, though view'd with a *Microscope* (which magnifies the Object, in bulk, above a million of times) yet I doubt not, but were we able *practically* to make *Microscopes* according to the *theory* of them, we might find hills, and dales, and pores, and a sufficient bredth, or expansion, to give all those parts elbow-room, even in the blunt top of the very Point of any of these so very sharp bodies.

Even something as mundane as the period at the end of a sentence could not escape Hooke's curiosity:

But leaving these Discoveries to future Industries, we shall proceed to add one Observation more of a *point* commonly so call'd, that is, the mark of a *full stop,* or *period.* And for this purpose I observed many both *printed* ones and *written;* and among multitudes I found *few* of them more *round* or *regular* then this ... but *very many* abundantly *more disfigur'd;* and for the most part if they seem'd equally round to the eye, I found those points that had been made by a *Copper-plate,* and Roll-press, to be as misshapen as those which had been made with *Types,* the most curious and smoothly *engraven strokes* and *points,* looking but as so many *furrows* and *holes,* and their *printed impressions,* but like *smutty daubings* on a matt or uneven floor with a blunt extinguisht brand or stick's end. And as for *points* made with a *pen* they were much *more ragged* and *deformed.*

In perhaps the most famous passage from *Micrographia*— and one that illustrates his attention to the details of the myriad of objects he studied under the magnification of his microscope, Hooke described the compound eye of a fly:

I took a large grey *Drone-Fly,* that had a large head, but a small and slender body in proportion to it, and cutting off its head, I fix'd it with the forepart or face upwards upon my Object Plate (this I made choice of rather than the head of a great blue Fly, because my enquiry being now about the eyes, I found this Fly to have, first the biggest clusters of eyes in proportion to his head, of any small kind of Fly that I have yet seen, it being somewhat inclining towards the make of the large *Dragon-Flies*. Next, because there is a greater variety in the knobs or balls of each cluster, then is of any small Fly.) Then examining it according to my usual manner, by varying the degrees of light, and altering its position to each kind of light, I drew that representation of it ... and found these things to be as plain and evident, as notable and pleasant.

Hooke goes on to describe in great detail exactly what he observed, including a count of the number of sections in which the eye was divided. He found that

the number of the *Pearls* or *Hemispheres* in the clusters of this Fly, was near 14000. which I judged by numbering certain rows of them several ways, and casting up the whole content, accounting each cluster to contain about seven thousand Pearls, three thousand of which were of a size, and consequently the rows not so thick, and the four thousand I accounted to be the number of the smaller Pearls next the feet and *proboscis*. Other Animals I observ'd to have yet a greater number, as the *Dragon-Fly* or *Adderbolt*: And others to have a much less company, as an *Ant*, &c. and several other small Flies and Insects.

By the time Pasteur and his contemporaries began thinking about the role of microbes in disease, western science — starting with Hooke and van Leeuwenhoek — had already developed a tradition of microscopic observations.

Even before the work of Louis Pasteur, others were instrumental in showing that microscopic organisms played a central role in disease. For instance, in the late nineteenth century the Hungarian physician Ignaz P. Semmelweis made the seemingly simple (and to modern observers, obvious) connection between hygiene and infection in maternity hospitals. The widespread practice among surgeons and physicians did not include simple cleaning of hands or instruments as they moved from patient to patient. Semmelweis argued that basic hygiene for both the patient and the physician drastically reduced infection rates.

In 1854, perhaps the most famous episode in epidemiology occurred when the English physician John Snow used careful observations along with statistical data to trace the source of a major cholera epidemic in London to a particular public water pump. Even though the epidemic diminished after the pump was closed, other physicians and scientists could not believe that microscopic organisms a tiny fraction of the size of a human could sicken and even kill. Eventually, though, resistance to the germ theory began to erode with the work of Pasteur his contemporaries.

After achieving success and fame with the new method of pasteurization, Louis Pasteur turned his attentions to vaccinations against deadly diseases. He and his team of researchers showed that weakened anthrax germs injected into sheep protected the sheep from contracting the deadly (and highly communicable) disease. Perhaps Pasteur's greatest triumph was his application of a rabies vaccine on a young boy recently bitten by rabid dog. The vaccine worked, most certainly saving the boy from an agonizing death. Although he did not discover the first vaccine — Edward Jenner, for instance, had developed a highly successful smallpox vaccine almost a century earlier — Pasteur's new vaccinations and his understanding of the underlying cause of disease and disease prevention laid the groundwork for the most important advance in modern medicine.

In "The Germ Theory and its applications to medicine and surgery," first read before the French Academy of Sciences in 1878, Pasteur explained the experimental results that led him to embrace the idea that microbes were the root cause of disease:

> The Sciences gain by mutual support. When, as the result of my first communications on the fermentations in 1857–1858, it appeared that the ferments, properly so-called, are living beings, that the germs of microscopic organisms abound in the surface of all objects, in the air and in water; that the theory of spontaneous generation is chimerical; that wines, beer, vinegar, the blood, urine and all the fluids of the body undergo none of their usual changes in pure air, both Medicine and Surgery received fresh stimulation. A French physician, Dr. Davaine, was fortunate in making the first application of these principles to Medicine, in 1863.
>
> Our researches of last year, left the etiology of the putrid disease, or septicemia, in a much less advanced condition than that of anthrax. We had demonstrated the probability that septicemia depends upon the presence and growth of a microscopic body, but the absolute proof of this important conclusion was not reached. To demonstrate experimentally that a microscopic organism actually is the cause of a disease and the agent of contagion, I know no other way, in the present state of Science, than to subject the *microbe* (the new and happy term introduced by M. Sedillot) to the method of cultivation out of the body. It may be noted that in twelve successive cultures, each one of only ten cubic centimeters volume, the original drop will be diluted as if placed in a volume of fluid equal to the total volume of the earth. It is just this form of test to which M. Joubert and I subjected the anthrax bacteridium. Having cultivated it a great number of times in a sterile fluid, each culture being started with a minute drop from the preceding, we then demonstrated that the product of the last culture was capable of further development and of acting in the animal tissues by producing anthrax with all its symptoms. Such is — as we believe — the indisputable proof that *anthrax is a bacterial disease.*
>
> If it is a terrifying thought that life is at the mercy of the multiplication of these minute bodies, it is a consoling hope that Science will not always remain powerless before such enemies, since for example at the very beginning of the study we find that simple exposure to air is sufficient at times to destroy them.

The sentiment expressed by Pasteur in the preceding paragraph ("it is a terrifying thought that life is at the mercy of the multiplication of these minute bodies") was an almost universal belief before the germ theory of disease gained acceptance. How, physicians asked, could a tiny microscopic body be responsible for so many deadly diseases in a body so many degrees of magnitude larger than themselves? Pasteur explained that his experiments (and those of other scientists just beginning to realize the power of germs) were designed to find

> the absolute proof that there actually exist transmissible, contagious, infectious diseases of which the cause lies essentially and solely in the presence of microscopic organisms. The proof that for at least some diseases, the conception of spontaneous virulence must be forever abandoned — as well as the idea of contagion and an infectious element suddenly originating in the bodies of men or animals and able to originate diseases which propagate themselves under identical forms: and all of those opinions fatal to medical progress, which have given rise to the gratuitous hypotheses of spontaneous generation, of albuminoid ferments, of hemiorganisms, of archebiosis, and many other conceptions without the least basis in observation.

The name of Louis Pasteur lent credibility to the germ theory of disease. Yet, when he read his seminal paper "The Germ Theory and Its Applications to Medicine and Surgery" before the French Academy in 1878 he was building on, and in many ways culminating, the work of other pioneering scientists and physicians. In fact, only two years earlier a German physician became the first to prove that a particular disease was caused by a specific microbe.

Robert Koch was a child prodigy who studied medicine at the University of Göttingen. While serving as the district medical officer in a rural, agricultural region of Germany, Koch became interested in a search for the cause of anthrax, a serious disease that afflicted both humans and livestock. In spite of limited resources and his isolation from modern laboratories and medical colleagues, Koch's careful experiments and observations eventually proved that anthrax was indeed caused by the bacterium *Bacillus anthracis*. In his work "The Etiology of Anthrax, Founded on the Course of Development of the Bacillus Anthracis," published in 1876, Koch began by acknowledging that he is not the first to claim that a microbe is to blame for the anthrax disease:

> Since the discovery of rod-shaped bodies in the blood of animals that died of anthrax, much effort has been devoted to establishing that these cause not only direct transmissions of the disease but also its sporadic occurrence — to establishing that these constitute the characteristic contagium of the disease. Recently, Davaine has been principally occupied with this task. Supported by numerous inoculation tests with fresh or dried blood containing these rods, he asserted that the rods were bacteria and that the disease could occur only when these rods from anthrax blood were present. He ascribed cases of anthrax among people or animals, for which no direct transmission could be demonstrated, to the spread of dried bacteria by air currents, insects, and the like. He discovered that dried bacteria could remain alive for a long time. This seemed to have completely explained the spread of anthrax.

Koch went on to explain some of the criticisms of Davaine's conclusions, and noted his own findings:

> I often had the opportunity to examine animals that died of anthrax, so I conducted experiments to dispel these obscurities. I quickly concluded that Davaine's theory of the dissemination of anthrax was only partially correct.

Koch proceeded to describe in great detail the experiments he performed in order to understand the nature of anthrax. In particular, he detailed his attempts to show that the microbe *Bacillus anthracis* was responsible for the disease.

> So that every interested person will be able to confirm my results, I have described my methods exactly — methods that were often achieved only by tiring and time-consuming attempts. Moreover, it is particularly significant that Professor Cohn, in response to my request, expended the necessary effort to examine in detail a series of preparations and experiments that I presented at the Breslau Institute for Plant Physiology. He confirmed in every detail my assertions about the development of *Bacillus anthracis*. For this he deserves my great appreciation.

Koch claimed that his experiment "is sufficient proof that spores of *Bacillus anthracis* cause anthrax, when introduced directly into body fluids." Koch described his experiment thusly:

> It has been claimed that the disease caused by inoculation with anthrax blood is identical with septicemia. This claim could be taken as an objection to my inoculations with decaying anthrax substances. To refute this objection, I frequently inoculated mice with decaying

blood from healthy animals and with decaying aqueous humor and vitreous humor that was free from bacilli. These animals nearly always remained healthy. Of twelve inoculated mice, only two died; these died a few days after inoculation. They had enlarged spleens, but their spleens, as well as their blood, were completely free from bacilli. Moreover, I also inoculated animals with decaying vitreous humor in which a species of bacillus had spontaneously developed that was very similar to *Bacillus anthracis*. The spores of the two species could not be distinguished by size or appearance, but the filaments of the vitreous humor bacilli were shorter and more clearly articulated. In spite of numerous attempts, my inoculations with these bacilli and with their spores never caused anthrax. Animals also remained healthy after inoculation with spores of hay-infusion bacilli culture by Professor Cohn. On the other hand, I often inoculated with spore masses that had been cultured in vitreous humor and that, as I had convinced myself with microscopic examination, were derived from entirely pure cultures of *Bacillus anthracis*. The inoculated animals invariably died of anthrax.

The results of these experiments, Koch maintained, were simple and obvious: "It follows that only one species of bacilli can generate this specific disease. Other inoculated schizophytes are either harmless or cause a completely different disease process."

Koch concluded his seminal paper on anthrax by calling for more research into the microbes that cause other diseases. He presented his experimental method with anthrax as a guide:

We must follow the guidance of similar etiologies that we understand. Only then will it be possible to determine the essence of the infectious diseases, which have horribly devastated the human race, and to find reliable means of protecting ourselves against them.

After his success in identifying the anthrax germ, Koch went on to study other diseases such as cholera and tuberculosis. He tried but failed to develop a vaccine for tuberculosis; however, in his failed experiments he noted that a red rash formed around the injected region of patients who already carried the disease. This observation let to the modern-day diagnostic used to determine the presence of tuberculosis in a human carrier. It was for his work with tuberculosis that Koch was awarded one of the earliest Nobel Prizes for Medicine in 1905.

Whereas Pasteur and Koch were experimentalists preparing careful laboratory tests for the germ theory of disease, a British physician by the name of Joseph Lister was in the field, revolutionizing how surgeons understood infection in the operating and recovery rooms.

After his medical education in London, Lister quickly acquired a stellar reputation as a physician and surgeon. He was appointed professor of medicine at Glasgow (Scotland) University. Lister was appalled at the filthy conditions he found in the hospital, a situation mirrored in hospitals around the world. Lister was convinced that simple hygiene would help reduce the high mortality rates for patients. More than that, however, Lister became one of the early proponents of the germ theory that was just being unveiled by Pasteur, Koch, and others. The young surgeon learned that London was using carbolic acid used to treat sewage for microbes and decided to try the chemical on patients during surgery.

Lister's results were immediate and astounding! Patients recovering from surgeries who once died from infections now survived. Yet, surprisingly, Lister's methods were not universally accepted. Many physicians were tradition-bound and slow to adopt new ideas. In addition, the carbolic acid was corrosive to skin and generally unpleasant to work with.

This resistance, however, did not prevent Lister from advancing this new method for saving lives. In his work, *Antiseptic principle of the practice of surgery* (published in 1867), Lister wrote:

> In the course of an extended investigation into the nature of inflammation, and the healthy and morbid conditions of the blood in relation to it, I arrived several years ago at the conclusion that the essential cause of suppuration in wounds is decomposition brought about by the influence of the atmosphere upon blood or serum retained within them, and, in the case of contused wounds, upon portions of tissue destroyed by the violence of the injury. To prevent the occurrence of suppuration with all its attendant risks was an object manifestly desirable, but till lately apparently unattainable, since it seemed hopeless to attempt to exclude the oxygen which was universally regarded as the agent by which putrefaction was effected. But when it had been shown by the researches of Pasteur that the septic properties of the atmosphere depended not on the oxygen, or any gaseous constituent, but on minute organisms suspended in it, which owed their energy to their vitality, it occurred to me that decomposition in the injured part might be avoided without excluding the air, by applying as a dressing some material capable of destroying the life of the floating particles. Upon this principle I have based a practice of which I will now attempt to give a short account.
>
> The material which I have employed is carbolic or phenic acid, a volatile organic compound, which appears to exercise a peculiarly destructive influence upon low forms of life, and hence is the most powerful antiseptic with which we are at present acquainted.

Lister proceeded to give the results of numerous surgeries in which he first packed the wounds with rags soaked in acid and continued to apply the acid to the area for some time after the surgery. One of the many examples Lister detailed was of a young boy whose arm was saved:

> I left behind me in Glasgow a boy, thirteen years of age, who, between three and four weeks previously, met with a most severe injury to the left arm, which he got entangled in a machine at a fair. There was a wound six inches long and three inches broad, and the skin was very extensively undermined beyond its limits, while the soft parts were generally so much lacerated that a pair of dressing forceps introduced at the wound and pushed directly inwards appeared beneath the skin at the opposite aspect of the limb. From this wound several tags of muscle were hanging, and among them was one consisting of about three inches of the triceps in almost its entire thickness; while the lower fragment of the bone, which was broken high up, was protruding four inches and a half, stripped of muscle, the skin being tucked in under it. Without the assistance of the antiseptic treatment, I should certainly have thought of nothing else but amputation at the shoulder-joint; but, as the radial pulse could be felt and the fingers had sensation, I did not hesitate to try to save the limb and adopted the plan of treatment above described, wrapping the arm from the shoulder to below the elbow in the antiseptic application, the whole interior of the wound, together with the protruding bone, having previously been freely treated with strong carbolic acid. About the tenth day, the discharge, which up to that time had been only sanious and serous, showed a slight admixture of slimy pus; and this increased till (a few days before I left) it amounted to about three drachms in twenty-four hours. But the boy continued as he had been after the second day, free from unfavorable symptoms, with pulse, tongue, appetite, and sleep natural and strength increasing, while the limb remained as it had been from the first, free from swelling, redness, or pain. I, therefore, persevered with the antiseptic dressing; and, before I left, the discharge was already somewhat less, while the bone was becoming firm. I think it likely that, in that boy's case, I should have found merely a superficial sore had I taken off all the dressings at the end of the three weeks; though, considering the extent of the injury, I

thought it prudent to let the month expire before disturbing the rag next the skin. But I feel sure that, if I had resorted to ordinary dressing when the pus first appeared, the progress of the case would have been exceedingly different.

Lister concluded by relating how the general health of the patients in his hospital ward improved drastically with the use of his new antiseptic practice:

Previously to its introduction the two large wards in which most of my cases of accident and of operation are treated were among the unhealthiest in the whole surgical division of the Glasgow Royal Infirmary, in consequence apparently of those wards being unfavorably placed with reference to the supply of fresh air; and I have felt ashamed when recording the results of my practice, to have so often to allude to hospital gangrene or pyaemia. It was interesting, though melancholy, to observe that whenever all or nearly all the beds contained cases with open sores, these grievous complications were pretty sure to show themselves; so that I came to welcome simple fractures, though in themselves of little interest either for myself or the students, because their presence diminished the proportion of open sores among the patients. But since the antiseptic treatment has been brought into full operation, and wounds and abscesses no longer poison the atmosphere with putrid exhalations, my wards, though in other respects under precisely the same circumstances as before, have completely changed their character; so that during the last nine months not a single instance of pyaemia, hospital gangrene, or erysipelas has occurred in them. As there appears to be no doubt regarding the cause of this change, the importance of the fact can hardly be exaggerated.

Although Pasteur, Koch, and Lister were all convinced that diseases were the result of microbes in the body, the germ theory of disease only slowly became accepted by their peers. As scientists and physicians began tying microscopic life to specific diseases, and as they discovered vaccinations against — and in the case of Alexander Fleming and the discovery of penicillin, cures for — various diseases, the germ theory transformed medicine on a scale few other theories or discoveries can match. The success of the germ theory opened a new age of discovery aimed at the very small. And on another front, scientists began searching for the explanation of how life — whether it be on the scale of human life or the microscopic scale of bacteria — replicated and reproduced. In other words, what is the mechanism that underlies Mendelian genetics?

DNA and the Secret of Life

How are characteristics passed from parent to offspring? What is the mechanism that allows for this mystery of inheritance? Philosophers and scientists have speculated for centuries, but it was not until the twentieth century that the secret was discovered and the role of the molecule deoxyribonucleic acid (DNA) proven. The story that dramatically unfolded reveals the human side of science in a way few episodes in history can.

Even after the discovery of the existence of DNA, it was ignored by researchers because most were convinced that protein molecules held the answer to inheritance. The Canadian born physician Oswald Avery however, showed that it was indeed DNA, not proteins, that carried the code of life. Avery's 1944 paper on the role of DNA as the fundamental genetic material sparked a race to uncover the structure of the DNA molecule, and with it perhaps the mechanism for genetic reproduction.

In early 1948, the American chemist Linus Pauling discovered the helical structure of proteins. This discovery, coupled with more experimental data from the early 1950s, formed the foundation on which a structure for DNA was built. In fact, Pauling and his son, Peter, led an American team racing to be the first to discover this structure. Although they were eventually beaten to the discovery by a team of scientists working in England, Linus Pauling's contributions to the understanding of the structure of chemical bonds and the helical structure of proteins led to a Nobel Prize in chemistry in 1954. As it turns out, Pauling's fame continued to rise in areas outside of chemistry. Pauling became one of the first outspoken advocates for the need for vitamin supplements, particularly vitamin C. Pauling was convinced of the efficacy of using large doses of vitamin C to prevent colds, the flu, and even cancer. Not content to sit on his scientific laurels, Linus Pauling also led the movement to ban atmospheric testing of nuclear weapons and to seek peaceful alternatives to a much-feared nuclear war. For his work, Pauling received the 1962 Nobel Peace Prize.

While Pauling was searching for clues to the structure of DNA in America, several scientists were conducting their own investigations in England. Maurice Wilkins, Rosalind Franklin, Francis Crick, and James Watson soon found themselves in a race the results of which changed the science of genetics.

James Watson was born in 1928 in Chicago into a middle class American family. He entered the University of Chicago at the tender age of 15, where he earned a bachelor's degree in zoology. He moved to the University of Indiana, where he became interested in genetics and received his Ph.D. The young scientist's interest in bacteriophages led him to the University of Copenhagen for a postdoctoral position. While attending a symposium in Naples, Watson had the good fortune of meeting Maurice Wilkins, a New Zealand-born (but London-based) physicist currently working on images of the DNA molecule acquired using a method called x-ray diffraction. Wilkins' work proved enticing enough for Watson that he was able to obtain for himself an invitation to work at the famed Cavendish Laboratory in Oxford — one of the epicenters of research into the DNA molecule. It was here that James Watson met Francis Crick, another young researcher with a strong interest in DNA.

Like James Watson, Francis Crick's (1916–2004) childhood was middle class and rather nondescript. After earning a degree in physics from University College, London, Crick's graduate studies were interrupted by the outbreak of World War II. During the war, Crick served as a scientist for the British Admiralty. After the war, the young physicist made an important career decision. Disenchanted with physics, and finding little success in the field, Crick decided to switch to molecular biology and began studying the use of x-ray crystallography in the study of the structure of proteins. Crick eventually found a home at the Cavendish Laboratory in 1949, where he began serious work on a Ph.D. thesis.

Late 1951 saw the beginning of one of the most important partnerships in the history of science. Soon after James Watson moved to the Cavendish Laboratory, he and Francis Crick agreed to work together on the mystery of the structure of the DNA molecule. This unlikely pair of an American zoologist and a British physicist embarked on a convoluted and controversial path of scientific discovery. The ultimate result laid the foundation for decades of advances in genetics.

Neither Watson nor Crick had originally come to Cambridge to work on the structure of DNA. However, their insatiable curiosity and unusual partnership guided them towards a race to understand the mechanism of cell replication through DNA. Before Watson arrived at Cambridge, Maurice Wilkins, a physicist and veteran of the Manhattan Project, and Rosalind Franklin, a chemist, were already studying the DNA molecule at Kings College in London using x-ray crystallography. Unfortunately, Wilkins and Franklin did not get along, and the intrigue that ensued represents a black mark on the history of science. In James Watson's classic personal account of the story of DNA, *The Double Helix*, he described his personal view of Franklin, usually in unflattering and unfriendly terms:

> I suspect that in the beginning Maurice hoped that Rosy would calm down. Yet mere inspection suggested that she would not easily bend. By choice she did not emphasize her feminine qualities. Though her features were strong, she was not unattractive and might have been quite stunning had she taken even a mild interest in clothes. This she did not. There was never lipstick to contrast with her straight black hair, while at the age of thirty-one her dresses showed all the imagination of English blue-stocking adolescents. So it was quite easy to imagine her the product of an unsatisfied mother who unduly stressed the desirability of professional careers that could save bright girls from marriages to dull men.... Clearly Rosy had to go or be put in her place.... The thought could not be avoided that the best home for a feminist was in another person's lab.

Watson related an almost comical incident that happened in Franklin's laboratory when they disagreed strongly over the interpretation of some x-ray data:

> Suddenly Rosy came from behind the lab bench that separated us and began moving toward me. Fearing that in her hot anger she might strike me, I grabbed up the Pauling manuscript and hastily retreated to the open door. My escape was blocked by Maurice, who, searching for me, had just then stuck his head through. While Maurice and Rosy looked at each other over my slouching figure, I lamely told Maurice that the conversation between Rosy and me was over and that I had been about to look for him in the tea room. Simultaneously I was inching my body from between them, leaving Maurice face to face with Rosy. Then, when Maurice failed to disengage himself immediately, I feared that out of politeness he would ask Rosy to join us for tea. Rosy, however, removed Maurice from his uncertainty by turning around and firmly shutting the door.
>
> Walking down the passage, I told Maurice how his unexpected appearance might have prevented Rosy from assaulting me. Slowly he assured me that this very well might have happened. Some months earlier she had made a similar lunge toward him. They had almost come to blows following an argument in his room. When he wanted to escape, Rosy had blocked the door and had moved out of the way only at the last moment. But then no third person was on hand.

In the Epilogue of *The Double Helix*, Watson softened his assessment of Franklin. After moving back to the United States, he saw much less of Franklin than did Crick, but noted a change in his assessment of her:

> By then all traces of our early bickering were forgotten, and we both came to appreciate greatly her personal honesty and generosity, realizing years too late the struggles that the intelligent woman faces to be accepted by a scientific world which often regards women as mere diversions from serious thinking. Rosalind's exemplary courage and integrity were apparent to all when, knowing she was mortally ill, she did not complain but continued working on a high level until a few weeks before her death.

The Double Helix was a controversial book. Rosalind Franklin was not the only fellow scientist portrayed by Watson in a less than flattering light, while Watson himself was often depicted as a heroic figure. In addition, Watson's account of the scientific method employed by he and Crick is much different than the textbook definition of the scientific method. Informal, disorganized, and occurring at many places outside of the laboratory (including the local pub), Watson's description of the events leading up to one of the most famous discoveries in the history of science does not paint the traditional picture of the detached and disinterested researcher working diligently, immune to interference from the outside world.

Reading *The Double Helix* often feels like reading a personal diary, as in when Watson revealed the instant he made a groundbreaking discovery about the structure of DNA:

> If this was [*the structure of*] DNA, I should create a bombshell by announcing its discovery. The existence of two intertwined chains with identical base sequences could not be a chance matter. Instead it would strongly suggest that one chain in each molecule had at some earlier stage served as the template for the synthesis of the other chain. Under this scheme, gene replication starts with the separation of its two identical chains. Then two new daughter strands are made on the two parental templates, thereby forming two DNA molecules identical to the original molecule.

Although Watson immediately noted several flaws in the details of his model, he remained cautiously optimistic that he and Crick could work out any problems that arose. In spite of the fact that Watson felt "my scheme was torn to shreds by the following noon," he and Crick continued to experiment with different versions of a double helix. Even with several advances in the days to follow, Watson admitted that he felt some trepidation when his partner somewhat prematurely announced success at the local pub which they and other Cavendish scientists frequented: "Thus I felt slightly queasy when at lunch Francis winged into the Eagle to tell everyone within hearing distance that we had found the secret of life."

After working out the finer details, checking their results against experimental data and several widely accepted theories, Watson and Crick sought, and immediately gained, the support for their model of several colleagues, including (much to Watson's surprise), Rosalind Franklin:

> Rosy's instant acceptance of our model at first amazed me. I had feared that her sharp, stubborn mind, caught in her self-made antihelical trap, might dig up irrelevant results that would foster uncertainty about the correctness of the double helix. Nonetheless, like almost everyone else, she saw the appeal of the base pairs and accepted the fact that the structure was too pretty not to be true. Moreover, even before she learned of our proposal, the X-ray evidence had been forcing more than she cared to admit toward a helical structure.

Immediately Watson and Crick began writing their historic paper, which appeared alongside papers supporting their conclusions with experimental evidence, co–authored by, among others, Wilkins and Franklin.

In the article entitled "A Structure for Deoxyribose Nucleic Acid," which appeared in the journal *Nature* on April 25, 1953, Watson and Crick began simply: "We wish to suggest a structure for the salt of deoxyribose nucleic acid (D.N.A.). This structure has novel features which are of considerable biological interest."

After pointing out the weaknesses of the structures proposed by Pauling and others,

Watson and Crick related: "We wish to put forward a radically different structure for the salt of deoxyribose nucleic acid. This structure has two helical chains each coiled round the same axis."

The authors then expanded on the details of their model. They pointed out,

The novel feature of the structure is the manner in which the two chains are held together by the purine and pyrimidine bases. The planes of the bases are perpendicular to the fibre axis. They are joined together in pairs, a single base from one chain being hydrogen-bonded to a single base from the other chain, so that the two lie side by side with identical z-coordinates. One of the pair must be a purine and the other a pyrimidine for the bonding to occur.

Watson and Crick admitted that their proposed model required more experimental proof before it could be accepted. They noted, however, that other papers published by their colleagues (including Wilkins and Franklin) in the same edition of *Nature* might supply some of this needed affirmation of their theory.

The previously published X-ray data on deoxyribose nucleic acid are insufficient for a rigorous test of our structure. So far as we can tell, it is roughly compatible with the experimental data, but it must be regarded as unproved until it has been checked against more exact results. Some of these are given in the following communications. We were not aware of the details of the results presented there when we devised our structure, which rests mainly though not entirely on published experimental data and stereo-chemical arguments.

Of course, some mention must be made of the obvious: the entire reason that the structure of DNA is of interest is due to its role in inheritance, a fact that Watson and Crick mentioned almost nonchalantly: "It has not escaped our notice that the specific pairing we have postulated immediately suggests a possible copying mechanism for the genetic material." The authors closed the short communication with a promise to publish more details forthwith, and mention of a few scientists to whom they are indebted.

Understanding the structure of DNA, and more importantly how this structure leads to replication of genetic material and thus inheritance, formed the foundation for genetics for the rest of the century. Watson, Crick, and Wilkins were awarded the Nobel Prize in physiology or medicine in 1962. Rosalind Franklin was never so honored, having passed away four years earlier (the Nobel committee does not award the prize posthumously). Both Watson and Crick became leaders in the scientific community. In fact James Watson became the first director of the Human Genome Project, an ambitious attempt to map the entire human genome completed in 2003.

It often seemed to his contemporaries that Pasteur was extremely fortunate in that many of his discoveries seemed particularly serendipitous. In fact, on the surface of things this seems to be the case for many famous scientists in history who were in the right place at the right time. In this chapter alone, we have witnessed Lister's opportune insight into the use of acid, routinely used to treat garbage, as a disinfectant for wounds; Koch's presence in the right place at the right time to discover the role of microbes in causing particular diseases; and the chance meeting of Watson and Crick at a time and place ripe for the discovery of the structure of DNA. Yet others were present at the same time and place. Why not them? Pasteur very eloquently and succinctly answered for them all: "Chance favors the prepared mind."

PRIMARY SOURCES

De Vries, Hugo. *Species and Varieties: Their Origin by Mutation*. Edited by Daniel Trembly MacDougal, 2d ed. Chicago: Open Court, 1906.

Koch, Robert. "The Etiology of Anthrax, Founded on the Course of Development of the Bacillus Anthracis." *Essays of Robert Koch*, translated by K. Codell Carter. New York: Greenwood Press, 1987.

Lister, Joseph. *Antiseptic Principle of the Practice of Surgery*. Modern History Sourcebook. http://www.fordham.edu/halsall/mod/modsbook.html.

Mendel, Gregor. *Experiments in Plant Hybridization*. http://www.mendelweb.org.

Pasteur, Louis. "The Germ Theory and Its Applications to Medicine and Surgery." *Scientific Papers. Physiology, Medicine, Surgery, Geology*, edited by Charles W. Eliot. The Harvard Classics, vol. 38. New York: P.F. Collier & Son, 1959.

Watson, James. *The Double Helix: A Personal Account of the Discovery of the Structure of DNA*. New York: Simon & Schuster, 1968.

Watson, J. and Crick, F. "A structure for Deoxyribose Nucleic Acid," *Nature* 171 (April 25, 1953).

OTHER SOURCES

Bowler, Peter J. *The Mendelian Revolution: The Emergence of Hereditarian Concepts in Modern Science and Society*. Baltimore: Johns Hopkins University Press, 1989.

Debré, P. *Louis Pasteur*. Baltimore: Johns Hopkins University Press, 1998.

Dubos, René J., and Thomas D. Brock. *Pasteur and Modern Science*. Madison, WI: Science Tech Publishers; Berlin and New York: Springer-Verlag, 1988.

Farmer, Laurence. *Master Surgeon: A Biography of Joseph Lister*. New York: Harper, 1962.

Gradmann, Christoph. *Laboratory Disease: Robert Koch's Medical Bacteriology*. Baltimore: Johns Hopkins University Press, 2009.

Henig, Robin Marantz. *The Monk in the Garden: The Lost and Found Genius of Gregor Mendel, the Father of Genetics*. Boston: Houghton Mifflin, 2000.

McElheny, Victor K. *Watson and DNA: Making a Scientific Revolution*. Cambridge, MA: Perseus, 2003.

Olby, Robert C. *Francis Crick: Hunter of Life's Secrets*. Cold Spring Harbor, NY: Cold Spring Harbor Laboratory Press, 2009.

_____. *The Path to the Double Helix*. London: Macmillan, 1974.

Orel, Vítezslav. *Mendel*. Oxford; New York: Oxford University Press, 1984.

Ridley, Matt. *Francis Crick: Discoverer of the Genetic Code*. New York: Atlas Books, 2006.

Strathern, Paul. *Crick, Watson, and DNA*. New York: Anchor, 1999.

Sturtevant, A. H. *A History of Genetics*. New York: Harper & Row, 1965.

Waller, John. *The Discovery of the Germ: Twenty Years That Transformed the Way We Think About Disease*. New York: Columbia University Press, 2002.

Index